Soft Shells

Books are to be returned on
the last date below

10 JAN 2003

Hans-Joachim Schock

Soft Shells
Design and Technology of Tensile Architecture

Birkhäuser
Basel · Berlin · Boston

This publication was kindly supported by
Festo KG, Esslingen, Germany and
Canobbio spa, Castelnuovo Scrivia, Italy

If existing patents, utility models or trademarks are not mentioned in this book, this is not to be taken as an indication that these do not exist for the products or tradenames.

Illustration on page 2:
Pneumatic hall 'Airtecture', Esslingen-Berkheim; pneumatic Y-shaped column and pneumatic tension elements ('muscles')

Library of Congress Cataloging-in-Publication Data
A CIP catalogue record for this book is available from the Library of Congress, Washington, D.C., USA

Die Deutsche Bibliothek - CIP Einheitsaufnahme
Schock, Hans-Joachim : Soft shells : design and technology of tensile architecture / Hans-Joachim Schock. – Basel ; Berlin ; Boston : Birkhäuser, 1997
Dt. Ausg. u.d.T.: Schock, Hans-Joachim : Segel, Folien und Membranen
ISBN 3-7643-5450-X (Basel ...)
ISBN 0-8176-5450-X (Boston ...)

This book is also available in a German language edition
(ISBN 3-7643-5449-6)

This work is subject to copyright. All rights are reserved, whether the whole or part of the material is concerned, specifically the rights of translation, reprinting, re-use of illustrations, recitation, broadcasting, reproduction on microfilms or in other ways, and strorage in data banks. For any kind of use permission of the copyright owner must be obtained.

© 1997 Birkhäuser – Verlag für Architektur, P.O.Box 133,
CH - 4010 Basel, Switzerland
Printed on acid-free paper produced from chlorine-free pulp.
TCF ∞
Design: Martin Schack, Dortmund
Printed in Germany
ISBN 3-7643-5450-X
ISBN 0-8176-5450-X

9 8 7 6 5 4 3 2 1

Contents

7 Introduction: Sails, Foils and Membranes

33 Temporary Library at UCLA, Los Angeles, USA

39 Amenities Building of the Inland Revenue Centre, Nottingham, Great Britain

43 Research Laboratory Venafro, Pozzilli, Italy

48 Recreational Clinic, Masserberg, Germany

54 Butterfly House, Berlin, Germany

58 Extension of Lycée la Bruyère, Versailles, France

62 Carnivore and Palm House, Munich Zoo, Germany

68 High-Rise Facade, Osaka, Japan

71 Dubai Creek Marina Club, Dubai

75 Ministerial Leisure Centre, Riyadh, Saudi Arabia

78 Covered Tennis Court, Gorle, Italy

82 Kaetsu Memorial Gymnasium, Tokyo, Japan

86 Festival Theatre for the International Eisteddfod, Llangollen, Wales

92 Mobile Theatre for Buddy Holly Musical, Hamburg, Germany

97 National Research Exhibition 'Heureka', Zurich, Switzerland

102 Pneumatic Hall 'Airtecture', Esslingen-Berkheim, Germany

106 Suntory Pavilion, Expo '85, Tsukuba, Japan

111 Trade Fair Stand for the 'Automechanika', Frankfort, Germany

114 MEDIADROME Mobile Air Hall for 'ZDF' TV, Germany

119 Exhibition Building 'Big Wave', Hiroshima, Japan

124	Roof of Government Stadium, Hong Kong
128	Convention Centre, San Diego, California, USA
132	Roof Canopy of Sainsbury's Supermarket, Plymouth, Great Britain
136	Service Station, Wanlin, Belgium
141	Airport, Salzburg, Austria
147	CAMP DE MART Auditorium Roof, Tarragona, Spain
153	Climate-Control Parasols for the Extension to the Prophet's Holy Mosque, Medina, Saudi Arabia
160	Mobile Building System for Exhibitions 'AIC-Tensoforma'
164	Roof over Archaeological Excavation Sites on Pianosa and in Desenzano sul Garda, Italy
168	Mobile Bandstand, New York, USA
172	Bibliography
173	Acknowledgements
174	Index

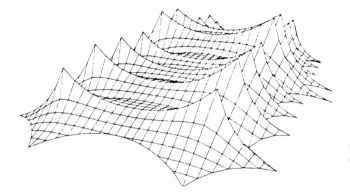

Wave surface: Airport buildings, Denver, Colorado (USA), isometric

Introduction: Sails, foils and membranes

Sails, foils and membranes – the mixture of heterogeneous terms in our title indicates the complexity, but above all the fascination of an extraordinary type of structure lasting from its beginnings until today. Lightweight, thin material, covering large spans without intermediate supports; an almost endless diversity of curved surface shapes; visually light, with thin edges, if desired transparent or translucent, and also physically light, quick and easy to erect.

At this point no summary of the history of membrane structures will be given and no glossary of membrane terminology will be delivered. We will not talk about the nomad tent, where everything began. Through Frei Otto membrane structures have advanced tremendously, first through his doctoral thesis published under the title "Hanging roofs"[1], then through his unusual and original tent structures, continuations of which are still emerging today from his Berlin office, the 'Entwicklungsstätte für den Leichtbau' (Developing Centre for Lightweight Structures), and later through his book "Zugbeanspruchte Konstruktionen" (Tensile Structures), which almost instantly became a kind of basic text book on this type of structures.

Then ultimately came the breakthrough, with the Olympic roof in Munich and its cable net structures, and with other large and wide-span roofs made from coated fabrics – for example by Walter Bird and his company Birdair, by David Geiger and Horst Berger and their (at first jointly led) office of Geiger Berger. Roofs made of coated fabrics became an established type of construction even for permanent roofs.

Some words on terminology: Membranes are thin 'skins', which can be made of fabric or foil or any other material. We all know what a fabrics is: it is made from warp and weft yarns, manufactured on a loom. High-performance man-made fibres give the necessary strength. Through coating the fabric can be waterproofed, thus also improving its durability, extending its service life and reducing its susceptibility to dirt accumulation. Foils on the other hand are made of thinly rolled or extruded homogeneous material, here above all from different types of plastics. Metal foils are rare. A sail, as we use the term here, is a type of construction similar to the sail on a sailing boat: a membrane surface with usually four corners, supported or fastened only at the corners and strengthened and held in between by a flexible cable along its edge. This is a special type of membrane structure.

With membrane structures two basic shapes can be distinguished, which we name using two terms from surface geometry:

Synclastic surfaces are curved like a sphere or a balloon for example. Anticlastic surfaces have their curvatures in opposite directions.

1 Otto, F.: Das hängende Dach (Hanging Roofs), Berlin; Bauweltverlag 1954

2 Otto, F. (editor): Zugbeanspruchte Konstruktionen, Frankfurt/M; Ullstein Verlag 1962

3 Berger, H.: Light Structures – Structures of Light, Basel, Berlin, Boston: Birkhäuser Verlag 1996

There the centres of the radii of the principal curvatures are situated on opposite sides of the surface. Examples are the saddle surface and the hyperbolic paraboloid.

Characteristic properties of membrane structures

The problems resulting from the Olympic roofs in Munich, which handicapped membrane structures in Germany for a long time, have fortunately become history. In the years after the construction of the Olympic roofs in Munich even small membrane structures were often re-

Sails: Yulara Holiday Resort, Alice Springs, Australia

Humped surface, membrane cladding on geodesic dome, humped supports using pneumatic cushions: Tennis hall, Essen, Germany

fused as being too expensive. Large, wide-span membrane roofs, like the airport building in Denver, Colorado (USA), and also ones with smaller or medium spans, as the Buddy Holly Musical Theatre in Hamburg for example, were chosen among other reasons for their economical solution for the building task, with their short production and construction period probably being the decisive advantage. A textile structure also offers the 'membrane advantage' of cost-saving possibilities in lighting and air conditioning, depending on location and requirements. Because of the translucence of the membrane artificial lighting is usually not necessary during the day, so that no energy needs to be absorbed and carried away by cooling. Air conditioning (i.e. cooling) is not imperative even in hot climates, since ventilation openings can be installed easily and a large portion of the thermal radiation is reflected by the roof especially in the case of one particular (and common) membrane material, PTFE-coated glass fibre fabric. There are a number of good, solid and practical reasons for using a textile structure, i.e. a membrane structure made from a modern, coated fabric material, apart from its appealing shape:

 – a short construction period and quick erection
 – their capability to form large column-free spaces
 – their relative economy together with the additional bonus of an attractive form
 – good heat protection during summer or in hot climates due to the high reflectivity of, for example, PTFE-coated glass fibre fabric
 – good earthquake resistance through their small mass.

Form

Equilibrium shape | Membrane structures behave in a fundamentally different way to the 'standard', i.e. stiff, conventional structures, we are familiar with. For stiff structures the visible form is not essentially dependent on the type and distribution of the loads – a truss keeps its form, independent of whether we load it with a distributed load or a single point load. Likewise a beam, joist or girder or a floor slab can be loaded perpendicular to its surface without difficulty.

For membrane structures other rules apply. In a membrane only tensile forces can be carried. Planar structures, plane-prestressed membranes can transfer no significant loads perpendicular to their plane; even under small loads large deformations and large membrane forces occur, which usually results in the system quickly becoming unusable.

Planar membranes are also susceptible to vibrations, which can be caused by the wind. To avoid these disadvantages, different remedies are available, turning membranes into useful structures:

3D-curvature, i.e. curving of the surfaces, so that loads can be carried in two independent directions due to the surface curvature. Together with prestress this creates a structure, in which virtually all distributed forces can be carried, even compression forces in arch direction, by reducing pre-tension.

High point surface: Outdoor and indoor swimming pool, Ettlingen-Schöllbronn, Germany

Arch surface: Cafeteria at Teachers Training College, Ludwigsburg, Germany

Pre-tensioning of the membrane, i.e. in the unloaded condition tension forces (tensile stresses) are present in the membrane in all directions, which under load can be reduced to zero if necessary.

Prestress and curvature of the membrane | Membranes are prestressed and receive a surface curvature to avoid flutter in the wind as well as pooling under snow load or rain. A slackening of the membrane is usually avoided by a sufficiently high prestress, while a short-term slackening under extreme loading conditions must not always be critical. With fabrics the prestressing forces and forces from loads are carried mainly in both directions of weave, i.e. in warp and weft, while with foils they act in the direction of the principal curvatures of the surface. Prestress and surface curvature are mutually dependent on each other; one being a function of the other through a mathematical relationship, connected via the equilibrium of forces, the ratio of the membrane radii and the one of the prestressing forces in the two directions, warp and weft.

Anticlastic shapes | With anticlastic shapes the centres of the radii of the principal curvatures lie on different sides of the surface. The tension forces from prestress or load are in equilibrium with each other, so that the surface is stabilised. In this case a prestressing of the membrane is necessary. A predetermined detail shape is, if possible at all, only feasible with a special prestress ratio, and vice versa not all shapes are possibly for a predetermined prestress ratio.

The anticlastically shaped membrane can be connected at the edge to a rigid or stiff substructure, for example to a steel girder or a concrete wall or slab. It can also have a flexible edge, for example an edge cable or an edge webbing, and in this case is only fastened point by point to a stiff substructure.

To achieve the necessary surface curvature, the membrane must be deformed out of its plane by support 'points'. For static reasons a point-shaped support is not possible in a mathematical sense, the support must instead be line-shaped or provide a finite bearing area, i.e. the 'point' must be enlarged into a line or area. In the case of a line support the membrane supported along a line formed by stiff structural elements, e.g. beams, arches or flexible elements, valley, ridge, or edge cables. The surface-shaped 'point' support is accomplished by cone-shaped, spherical or umbrella-like supports, cable loops or rosettes.

Despite the endless formal diversity of the membrane surfaces we may distinguish and define a limited number of basic types for practical applications of anticlastic membrane structures as roof structures:

Four-point surface, sails | This is the simplest membrane surface over a mostly square or rhombic plan, with four corners not in one plane. The edges of sails are usually built as cable edges; stiff edges, however, are also possible and meaningful.

Membrane surface as cladding: Supermarket canopy, Plymouth, England

Synclastic pneumatic structure: Miramar swimming pool, Weinheim, Germany

As a shape they are related (but not equal and limited) to the hyperbolic paraboloid (HP surface), which is obvious for structures with stiff edges.

Wave surfaces | The membrane surface stretches in undulatory fashion between high and low line supports, usually ridge and valley cables. With the 'parallel wave' the line supports proceed in plan parallel to each other, with the 'cross wave' (four high and four low points) and the 'star wave' (more than four low and high points) they radiate from a common centre.

Arch surfaces | Here the membrane surface is deformed out of its plane by one or several (compression) arches or arched frames.

High point surfaces, humped surfaces | Within the plan area one or several high points are arranged. The high points consist mostly of steel rings or umbrella-, paddle- or carpetbeater- (rosette-) shaped details, over which the membrane proceeds and where it is connected, – mostly by clamping. The high points are carried either by masts or they are suspended from a primary structure by cables. If the membrane runs over an umbrella or a humped – mostly curved – structure, we call it a 'humped surface'.

Membrane surfaces as decking or cladding | On a more or less conventional substructure, dome, grid structure or space frame the space-enclosing element is formed by a coated fabric membrane. Here some outstanding examples have been built in the last few years, including, for example, the supermarket canopy in Plymouth (Peter Rice, Ove Arup & Partner, cf. p. 132) and various gymnasium in Japan, such as the Amagi Dome and the Orio Gymnasium (cf. p. 25).

Synclastic shapes | With synclastic membrane shapes the centres of the radii of the principal curvatures lie on the same side of the surface. The tensile forces of the membrane are in equilibrium with the inside pressure (inflation pressure); acting together both stabilise the surface. Individual supports here are not customary, as the whole surface is prestressed and supported by the internal pressure; the same applies for the load introduction at supports i.e. foundations. The detailing principles described above apply analogously for suspending loads from the membrane, while altogether only lightweight loads can be suspended. Basic types of synclastic membranes are largely defined by their geometry:

Spherical shapes, cylinder shapes, for example standard airhalls, torus shapes and as a special case the quiltlike shapes of cable-reinforced stadium roofs. In addition there is an almost endless number of free shapes possible, which have, however, hardly any practical significance.

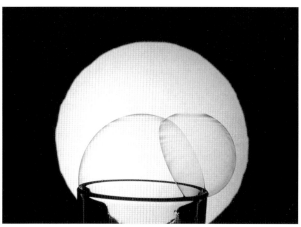

Synclastic pneumatic structure: MEDIA-DROM mobile airhall for ZDF TV, Germany

Synclastic pneumatic structure: Soap bubble

Material

Membranes can consist of different materials, of which every one has its specific possibilities and limits: coated fabric, uncoated fabrics, foils as well as different special types and variants.

Coated fabrics | Coated fabrics are the most frequently used membrane materials; here a base fabric of high-strength fibre is coated on both sides, whereby the fibres are protected from the environment, UV-radiation (polyester fabrics) or moisture (glass fibre fabrics). At the same time the membrane is made waterproof by the coating. PVC-coated polyester fabric is a standard material for small and large spans, with a limited service life, however, compared with conventional materials. By an additional surface coating PVC-coated polyester fabrics can be protected from pollution, whereby their aesthetics and possibly also the practical service life is increased. Such surface coatings for example are acrylic lacquers, PVDF coatings (e.g. Fluortop® by Ferrari) or Tedlar®-laminates.

PTFE-coated glass fibre fabric is also a standard material for wide span structures, where a high strength and service life comparable with conventional building materials is required.

PVC-coated glass fibre fabric has an improved fire resistance compared with PVC polyester fabric. Silicone-coated glass fibre fabric has improved flexibility and fire resistance. Various other coated fabric materials also exist, such as Neoprene-coated polyamide fabric, coated Aramide fabric with high strength and various other materials of less significance.

Uncoated fabrics | They are surely the oldest tent-building materials: very flexible, with a limited watertightness, however, and with limited strength and service life. They are mainly used for smaller membrane structures, and through their high flexibility they are also well suited for variable (movable) roofs with smaller spans. The most common material in this category are cotton- and cotton-mix fabrics, acrylic fabrics and, as a recent development, uncoated PTFE-fabrics.

Foils | All foils are characterised by their large shear stiffness, requiring particular care at the pattern cutting stage and possibly also particular patterning methods.

They also have in common a high transparency and translucence, which so far is not achieved by any other membrane material. While the service life of PVC-foil is limited, it is rather good for ETFE foil.

Nets | Wire fabrics, man-made fibre nets and steel cable nets are a special form of membrane. They can be used with (e.g. the Olympic roofs in Munich) and without infill panels or cladding (e.g. aviaries).

Shade fabrics | A special type, situated somewhere between net and PVC-coated polyester fabric are shade fabrics. Here a net fabric with a mesh width of some millimetres is manufactured in a special weaving process and coated with PVC to improve its durability and shading effect. These fabrics are porous, permeable to air and obviously not waterproof. By coating the fabric with foam the shading effect can be improved.

Grid fabrics | Another special type are grid fabric. Here a net fabric (usually polyester), produced in a special weaving process is coated with a highly translucent coating (usually PVC), resulting in a high translucency and usually some reduction of the service life.

Rope edge

Edge-rope sleeve | At the edge the membrane surface must either be connected to a rigid structural component or strengthened and held with a flexible edge reinforcement, such as webbing, cable or a combination of the two.

The edge cable can run in a sleeve at the membrane edge. The membrane is cut with an edge-seam strip, turned over and welded together, which due to the edge cable curvature results in wedge-shaped openings which are either left open or are closed by an additional reinforcement strip. Frequently the membrane edge is cut with a small edge overstand and has a diagonally cut (bias cut) sleeve welded on. As reinforcement in longitudinal direction and for the transfer of tangential forces into the corner plates an additional webbing along the edge may be required.

Garland edge | Instead of cutting and connecting the edge cable at every corner it may be designed as a garland cable. Here a continuous edge cable runs in a small cable sleeve past several corner plates. The corner plates are clamped onto the membrane from above and below. This detail is used as a simple solution for light membranes, but also for large membrane forces where the edge cable forces and dimensions are correspondingly greater. In the latter case a lightweight garland cable is introduced. It has short spans and is connected pointwise along the large edge cable, supporting it in its function.

Clamped edge | When using stiff fabrics with little tolerance against cutting pattern inaccuracies (like for example PTFE-coated glass fibre fabric) or for the higher (and stronger) grades of PVC-coated polyester fabrics, the membrane can be held in a clamping section with a boltrope, connected via fixed or adjustable connecting elements (e.g. turnbuckles) to the edge cable. Thus, slight patterning and material inaccuracies can be equalised by individual adjustments.

Gutters | Gutters can be constructed as conventional (for example sheet metal) gutters, positioned under the membrane edge and following their geometry. Loose, unstressed membrane aprons fastened to the membrane edge can be laid into the gutter and clamped there. Membrane gutters and special rain shields can also be connected to the membrane.
Here two typical solutions exist.

The rain bulge: A round section of foam rubber is inserted into a membrane sleeve and welded to the membrane or a plastic or metal angle is clamped along the edge. In light rainfall the bulge functions as a gutter shedding the rainwater to the side, in heavy rain it is flooded and overflows.

The membrane gutter: Short tube sections are inserted in membrane sleeves which have openings cut at the top, and welded onto the roof membrane. The tube pieces hold the gutter open, the water flows in through the top openings and away through the tubes to the sides.

Membrane aprons, wall connections | Membrane aprons form a connection between the membrane roof and a rigid wall or facade construction. They are usually not prestressed, but are held

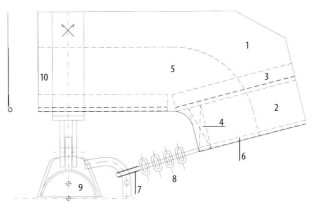
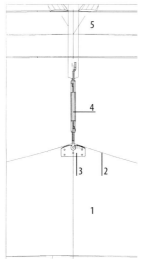

Edge cable sleeve 1:15: Ceremonial walkway cover at King's Court, Rabat, Morocco; 1 PVC - polyester membrane, 2 edge cable sleeve, 3 HF - weld, 4 reinforcement seam, 5 membrane corner reinforcement, 6 edge cable, 7 thimble, 8 cable clamp connection for simple length adjustment during installation, 9 steel substructure, 10 membrane passing over and supported on steel tube substructure.

Garland edge 1:50: Mobile theatre for Buddy Holly Musical, Hamburg, Germany; 1 membrane, 2 continuous edge cable in edge cable sleeve, 3 corner plates, 4 turnbuckle, 5 steel girder structures

along the edge through a clamping strip or formed by a membrane piece overshooting the edge. Welded-on aprons are problematic because of the peeling stresses they cause perpendicular to the weld. For slight movements of the roof membrane perpendicular to the edge a seal with a foam rubber bulge is also possible.

Firm edge

Clamping plates, clamped connections | When not using a prestressing device the connection of the membrane along a stiff edge can be made by means of bolt-rope and clamping strip. As the membrane needs to be prestressed in some way prestressing devices are necessary elsewhere. Other possible connections consist of tubes, bars or rods inserted into an edge sleeve, held with bolts or fixed to an edge tube via short cable loops.

Prestressing devices at the firm edge | Prestressing devices at the edge connections do not usually serve directly for applying a prestressing force, but as adjustment to equalise manufacturing tolerances. The prestressing is usually either applied by jacking up the masts using hydraulic jacks or by way of installing the guys with hydraulic jacks or tirfors under a predetermined prestress. At a stiff edge the prestress may be applied using a sheet metal or rolled steel gutter; here the prestress is applied using bolts or threaded bars. To avoid the necessity of having to re-cut galvanised threads, loose prestressing bolts are often used, inserted into steel shoes. The application of a series of turnbuckles in a row along a clamped edge (with possibly a gutter underneath) is equally possibly.

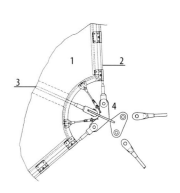

Corner detail with clamping plate edge: Amenities building at The Inland Revenue Centre, Nottingham, UK;
1 membrane, 2 edge cable with clamping plate edge and connecting plates, 3 connecting cable, 4 welded corner plate assembly

Membrane corners, corner points and corner plates | At membrane corners the forces from membrane, edge cable and edge webbing are collected and introduced into a corner plate, connected to a substructure. Here two basic solutions exist: At an open membrane corner, the membrane is cut back, only edge cables and webbing are connected to the corner plate; at a closed membrane corner the corner plate is clamped onto the membrane from above and below.

Point and area support | At high and low points the anticlastic membrane is deformed up and down perpendicular to a medium plane. As the detailing of true point supports is not possible because of the resulting high membrane forces, tending toward infinity, these 'points' must be enlarged into ring or surface-shaped supports.

Rings | High point rings can be suspended freely, i.e. via short cables connected to a mast or other primary structure or connected rigidly to the structure. For PVC-coated polyester fabric usually a cylindrical ring made from steel flat is used; for PTFE-coated glass fibre fabric the ring is mostly conical, its shape corresponding approximately with the incident angle of the mem-

Edge cable sleeve with edge webbing during manufacture; with the inserted chords the edge cable is pulled through during installation.

Membrane corner detail with clamping plate edge: Auditorium roof CAMP DE MART, Tarragona, Spain.

brane. The conical ring is welded up from sickle shaped plate strips, representing the shape of the developed cone surface, and is thus more costly.

High point rings can be closed by different cover structures: Steel boiler bottoms (obtainable as prefabricated steel components), light domes or structures of glass-fibre reinforced polyester resin (GRP), acrylic glass or polycarbonate or alternatively through a non-structural fabric membrane, made with a special cutting pattern and a separate prestressing or adjustment device.

Surface support, umbrella | As an alternative a membrane running over high or low points can be supported on surface areas. Common here are especially umbrella-shaped structures with rigid or flexible spokes, with their geometry usually following a spherical surface. The membrane runs across the spokes, it can be strengthened by a lining membrane and located by a bolt at the umbrella centre. The umbrella spokes can be widened by plywood strips. For smaller spans and forces hemispherical supports and plane circular plates have also been built, on top of which the membrane has to be formed, i.e. hot-air welded over a mould.

Line shaped support

Arch support | The membrane runs across an arch. Whether a lateral connection is made at the arch depends on the type and size of the structure and on the magnitude of the differential forces acting perpendicular to the arch plane, resulting from different loads on the adjacent membrane panels on both sides of the arch, e.g. due to snowdrift, or from different membrane spans. For moderate differential forces a connection between membrane and arch can be made by using a membrane apron under the arch. Alternatively the membrane can be fastened intermittently to the arch via garland cables, in which case space closure and watertightness again will be accomplished by non-structural membrane aprons clamped onto the arch or connected with a zipper section. Along the arches the membrane must also be prestressed, either manually or by a prestressing device at the arch ends (cf. p. 61).

Valley cables, ridge cables | Valley and ridge cables are usually placed directly onto the membrane which is strengthened on site. To equalise manufacturing tolerances an independent prestressing or adjusting device may be useful.

Eye cables | Cable loops lying in the membrane plane are called 'eyes'. Large eyes can be made like cable edges; the tighter curvature may, however, cause some additional complications. Small eyes and rosettes are rare because of the difficulties in making them.

Guys | Guys can be made as solid bars or cables. The tensile forces of the guys are introduced and transferred via steel parts into the gravity foundations, tension anchors or tension piles, usually together with a built-in prestressing device.

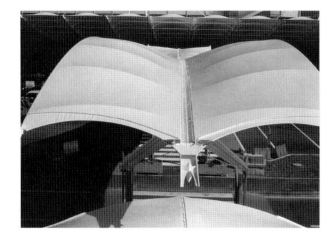
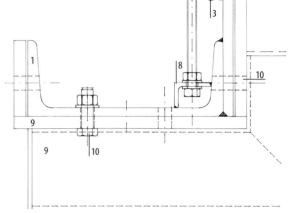

Guys without masts are equally possible. There the force direction does not change; the guy force proceeds in direction resulting from both edge cables and is thus usually quite flat, so that the incident point on the ground is generally situated far outside. This somehow limits the application of this type because of its large space requirement.

Prestressing devices

The following prestressing devices have been used and may be considered standard:

Ties as solid bars with left- and right-hand threads, screwed into a female thread at the joint (for prestressing the whole tie is turned) or with a special sleeve (similar to a turnbuckle).

Steel cables with end fittings and prestressing device (turnbuckle). A fitting is a steel connection element swaged into the cable end, mostly shaped as a fork or threaded fitting.

Threaded bars, anchored by casting them into the foundation, as a prestressing device for a single cable. The cable anchorage is accomplished with a swaged cable terminal on a steel transom, i.e. on a short, beam-like steel part between the threaded bars, and anchored there with a nut and counter nut. At the centre of the transom the guy cable is anchored by a swaged terminal, i.e. by a steel component swaged into the cable end.

Anchor bolts, cast into the concrete, are placed with a template to reduce tolerances in bolt spacing. The anchorage in the concrete is achieved by cast-in anchor bodies, the manufacturing and prestressing tolerances are taken up in a mortar joint, angular tolerances (around the bearing bolt axis) in the bolt. For this purpose convex (turned on a lathe to receive a cigar shape) or concave bolts to allow for angular tolerances in the plane of the bolt have also been used.

Corrosion protection

Due to movements in bolt connections resulting from varying loads and temperature changes local rusting can emerge, spoiling the appearance and reducing the service life of this connection. Caulking the connection details with (acid-free) silicone sealing compound prevents or reduces the penetration of moisture.

Prestressing gutter with lacing to a tube fixed in the gutter: Airport Salzburg, Austria

Prestressing gutter 1:15; Reconstruction of the Medical Academy 'Carl Gustav Carus', Dresden, Germany: 1 prestressing gutter, 2 roof membrane, 3 boltrope, 4 clamping strip, 5 clamping strip, 6 girder, 7 prestressing bolts, 8 steel angle with open slotted hole to receive bolt head, 9 gutter holder, 10 gutter connection

Some remarks on the projects

The projects presented here are a selection from material amply provided by the project authors, architects, planners, engineering practices and contractors. The projects all date approximately from the last ten years, and thus document current technology and its development. A further selection criterion, apart obviously from the quality of the building and the completeness of the material, was also variety in terms of type of construction, function and geographic location. Many excellent and fascinating projects could not be included, simply because of the size limitations of the book. I apologise to all those whose outstanding and fascinating projects have not been accepted only for these reasons.

I would also like to thank everyone who made available material, texts, photographs, drawings and much background information which was often specially produced and documented for this book, and also for their patience in answering my insistant questions and for having always reacted positively to my urging, despite pressures from imminent deadlines and their daily work. There are too many to name them all. Particular thanks also to the photographers of the splendid pictures, who gave their permission for publication.

Opposite page:

Climate control parasols for the extension of the Prophet's Holy Mosque, Medina, Saudi Arabia, view of courtyard with parasols in open position

National research exhibition 'Heureka', Zurich, Switzerland, aerial view with entrance tent and Galileitower

Recreational clinic, Masserberg, Germany, aerial View

Festival Theatre for the international Eisteddfod, Llangollen, Wales, membrane roof, view of permanent arena with main and edge arch

Side view with conventional wing buildings and permanent arena with main arch in the background

Auditorium roof CAMP DE MART, Tarragona, Spain, view from road

Pneumatic hall 'Airtecture', Esslingen-Berkheim, Germany, view

Interior; wall elements, transparent foil cushions and roof structure with translucent low pressure intermediate membranes

Bracing structure with y-column and pneumatic 'muscles'

Orio Gymnasium, Kita Kyushu City, Japan, 1989; interior views

Service station, Wanlin, Belgium, detail and view

Research Laboratory Venafro, Pozzilli, Italy, detail

Research Laboratory Venafro, Pozzilli, Italy,
view in twilight

Research Laboratory Venafro, Pozzilli, Italy, interior

Carnivore and palm tree house, Munich Zoo, Germany, foil cushion roof, interior

UCLA Towell Library, Los Angeles,
California, USA

View from south-east

Temporary Library at UCLA, Los Angeles, USA

This building is an unusual temporary library; it serves as accommodation for the collection of the main undergraduate library of the University of California in Los Angeles during work on the seismic upgrading of the existing library building. Despite its architectural impudence and freshness the 'Towell' (Temporary Powell Library) recalls shapes of the English Gothic: the round reading rooms at the end of the main library are reminiscent of late Gothic ribbed vaults, including a central ribbed shaft, such as can be found in the chapterhouses of English cathedrals. But the Towell is modern, provisional and not timeless: the ribs of the reading room are aluminium extrusions instead of stone; the windows are polycarbonate sheets instead of stained glass. And at the roof edge of the 'chapterhouse' the ribs curve out playfully like on a carrousel at the fairground.

The exuberant metaphorical imagery is the result of what the architect calls 'very opportunistic conduct on a severe budget'. Functional criteria, availability and construction period have determined this architecture. But this does not show.

Design

The design is based on an arrangement of membrane-covered bodies, which on one side take up the direction of the western campus axis and redirect it to the south, towards the busy Student Centre there. The temporary library is located on the UCLA campus in direct proximity to the old library, between two Romanesque revival buildings at the end of the Janns Memorial Stairway. The geometry and monumental classical balustrade of the stairway and the internal axes and radii of the different buildings were all integrated into the design of the temporary library. Thus the semi-circular, eastern reading room is arranged tangentially to the staircase, while the circular western reading room proceeds in the direction of the balustrade. According to the architects environmental advancement includes the use of recyclable components: mechanical and structural components can be dismantled and reused. Societal advancement involves the notion of bringing architectural value to functional, temporary structures. The development of a temporary building with an architectural discipline demonstrates that the humanistic aspects of architecture are not determined by the permanence of a given project.

Brief

The design programme was intended to provide, for a period of two-and-a-half years, a provisional home for the approx. 210 000 books, the special collections and library functions of the

Client
University of California

Architects
General and membrane structure: Hodgetts & Fung Design, Santa Monica, California, USA. Project partner: Craig Hodgetts, Hsin-Ming Fung. Staff members: Lynn Batsch, Robert Flock, William Martin jr., Peter Nobles

Structural Engineering
Membrane: Rubb Building Systems, Inc.
Structure and substructure: Robert Englekirk, Inc. and A. C. Martin & Associates

Specialist consultants
HVAC: The Sullivan Partnership
Electrical: Michael Feinberg, Patrick Byrne & Associates
Lighting: Patrick B. Quigley & Associates, Inc.

Contractor
Main contractor: American Constructors California, Inc.
Subcontractor for membrane: Rubb Building Systems, Inc.

Completion
1992

Facade section at the western reading room

Nylon chord connection at the facade and balustrade of the staircase at the western reading room

Powell Library of the University of California. Planning and construction of the 3300 m² large temporary library had to be completed within nine months, before in fall 1992 the new trimester began after the summer break. With a required minimum design life of five years it subsequently also had to accommodate other university departments, giving the 'Towell' a longer lifespan than some of the functions it houses.

The university wanted the building to preserve an existing plaza between the historic dance building and the men's gym. The siting also integrated existing components of the surrounding context and the foundations were designed to preserve the existing paving beneath the building.

Program

The ensemble consists essentially of four parts:

The central building with bookshelves on two floors, on the ground floor and on a mezzanine in the higher part of the building, with a periodicals and reference library, reading rooms, study carrels, a lecture hall, computer rooms and toilets.

The administrative wing which also houses the compact shelving.

The eastern and the western reading rooms.

Reading areas out in the open.

The main entrance is over ramps from the south-east. The subdivision into four separate buildings reduces the mass of the building volume and increases the flexibility for other usage.

Key plan 1:4000

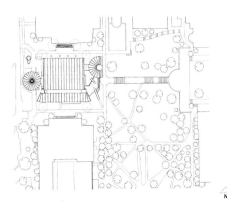

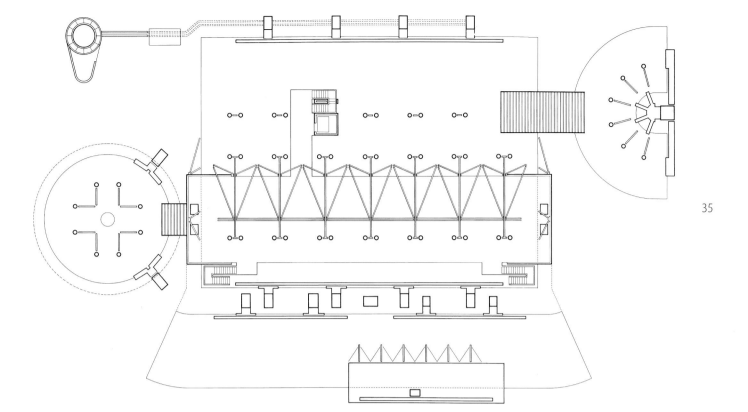

Plan 1st floor 1:500

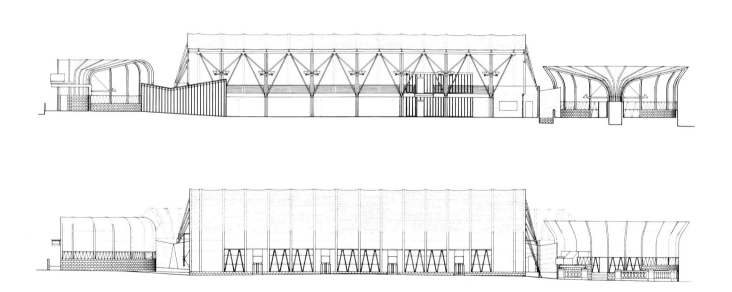

Longitudinal section, elevation 1:500

Structure | Since the project was subject to severe time and budget constraints, a re-usable aluminium and membrane roofing system was chosen for the main building with a short construction period, combined with a varied ensemble of disposable masonry, wood and plastic substructures. An integrally coloured concrete block, similar in hue to the brick and limestone typical of the campus architecture, suggests the foundations of a building which is no longer there and reinforces by contrast the temporary nature of the structure. The upper parts of the buildings, the mezzanines, are carried by a steel structure braced by compression struts; the ribs are braced horizontally in the middle of their span. Like in an old aircraft the structural fixings and fasteners necessary for the working of the building, steel cables, the visible load bearing elements and their working together are exposed and emphasised.

Interior view; looking toward the western reading room

The main building has a flat, shallow roof, closed at both longitudinal sides by cylindrical roof vaults. The roof over the administrative wing consists in the central part of a straight cylinder made of circular ribs of aluminium extrusions with an infill membrane of pre-tensioned PVC-coated polyester fabric. The end walls are turned slightly toward the main building resulting in torus-shaped transition parts. The cylinder is penetrated on the exterior side by the flat roof of the office wing covered with corrugated roofing panels. The connecting part between main building and administration has a flat roof and houses the toilets. The roof has an earthquake-resistant structure of aluminium extrusions, a bracing of rectangular steel tubes and a cable suspension in the arch area.

Inside there are V-shaped struts forming the earthquake X-bracing with tension diagonals and turnbuckles with tubular steel struts in the roof plane. The facade is braced by eccentrically connected struts in and out of its plane visible from the outside. Flexible details at the connection of the roof structure with the gable walls were developed to accommodate the extreme lateral movements under wind and especially under earthquake loads and to effectively separate roof and gable walls seismically.

The mezzanine is supported by columns made from steel I-sections and square hollow sections; the external walls consist partly of wood or metal stud construction on a base of block

Interior view of the entrance bridge on the 1st floor

masonry on concrete strip foundations. The steel detailing is functional and simple, the connections welded or bolted, and equipped with steel gusset plates.

Foundations | The foundations are conventional reinforced concrete strip foundations and ground slabs, with the steel structure and aluminium ribs bolted on directly.

Membrane | The translucent roof membrane is composed of a pre-tensioned PVC-coated polyester fabric on aluminium extrusion ribs with I-section and a groove to receive the membrane anchorage. The roof membrane is held at its edge by a boltrope running in an aluminium edge profile and is tensioned from the other side by a square head bolt together with a tension nut. An aluminium cap protects against the weather. For the roof membrane the standard fabric was modified to satisfy the special requirements of thermal insulation and sound absorption. For this a 6mm-thick foam backing was glued onto the inside the roof membrane. The western reading room has a nylon chord fabric-to-wall connection with an insulation foam backing behind.

Construction and equipment

In the lower part the walls are clad with a curtain wall facade of corrugated GRP sheets on wood or metal stud wall construction, with an OSB panelling and facing brick masonry in the base and are glazed in the upper area of the end walls with polycarbonate glazing on metal ribs.

The building parts were joined by connecting passages furnished with easy chairs and through a service block with toilets and plant rooms.

The air conditioning is achieved with outside air-conditioning units, 21 heat pumps and a cooling tower connected with the heating and cooling system of the University. All ductwork is exposed and partially serves as wall infill. The interior is protected against fire by a sprinkler system. Natural lighting is achieved through transparent plastic sheet roof lights and through glazed gable walls. The glazing consists of flexibly mounted polycarbonate sheets, to accommodate the movements of the building.

In general, available standard components and functional detailing were used, which, together with the exposed technical plant, result in a functional design.

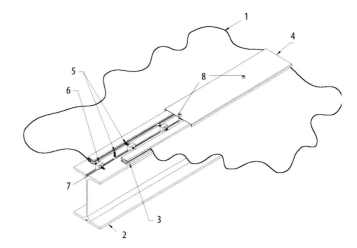

Structure, pre-stressing detail:
1 roof membrane, PVC-coated polyester fabric with foam backing and welded-on boltrope, 2 aluminium extrusion, I-section with grooves to receive square head bolts, 3 boltrope section, aluminium extrusion, 4 aluminium cap, 5 square head bolt, 6 aluminium tensioning block, 7 tension nut, 8 fixing nut for cap

Amenities Building of the Inland Revenue Centre, Nottingham, Great Britain

When the Inland Revenue decided in 1989 to transfer much of their work away from London the site selected as a new location was the Castle Meadow area in Nottingham. The disused industrial site was ideal for an office development. Pile foundations for a design-and-build scheme were already in place when in March 1991 work was suspended following criticism from the Royal Fine Arts Commission and other institutions. A new design was developed through a staged architectural competition which was won by the architects Michael Hopkins & Partner.

The winning scheme

The competition brief called for a scheme designed according to cost-benefit criteria, which would be flexible enough to accommodate changed demands of the future, including another occupier, i.e. it should be suitable for rental. A 'green', ecological design with maximum use of natural light and ventilation was desired, also a short construction period and a budget limited to £ 50 million. The view of the castle also had to remain unobstructed.

The complex is composed of four L-shaped buildings and two with a courtyard as well as a central amenities building entirely different in design. The quality and fascination of the design lies in the combination of an innovative and optimized traditional construction with the unusual structure of the central amenities building with its membrane roof.

The office buildings with 4000 m^2 office area are marked visually through the structural brick piers and the stair towers at the corners. The wish for natural ventilation led to a building depth of only 13.6 m and a limited building height of three and four floors. The roofs are lead-covered on plywood panels. The facade reduces the solar gain to a minimum, while the building mass stores heat during the day. At night the building mass is cooled by ventilation contributing to a comfortable room climate the next day. The ventilation concept is relatively simple; air enters either through room-high sliding windows or through low-level, fan-assisted ventilation grilles and is extracted through the cylindrical stair towers acting as solar chimneys. For this purpose the stair towers are equipped with lightweight membrane roofs, which can be lifted.

The amenities building

It contains the staff restaurant, sports facilities and a crèche. The building has the form of two slightly curved, double-storey blocks, which enclose the central sports area. The complex is

Client
The Inland Revenue

Architects
Michael Hopkins & Partner, London

Structural engineering
General, membrane, structure and substructure: Ove Arup & Partners, London

Specialist consultants
HVAC: Ove Arup & Partners
Quantity Surveyors: Turner and Townsend Quantity Surveyors
Project management: Turner and Townsend Quantity Surveyors

Contractors
Project management:
Laing Management
Membrane & steel:
KOIT High-Tex GmbH

Completion
1995

Aerial view of IRS complex, with castle in the background

Office buildings with cylindrical stair tower as solar chimney with circular membrane lifting roof

closed at one end by the reception area and on the other by the kindergarten. Both ends are fully glazed. The membrane roof as a unifying feature covers the whole building. The membrane edges remain visible, and the central membrane is separate from the one over the side blocks, thus emphasising the lightness of the roof structure. The discontinuity not only emphasises the different functions, but also breaks up the mass of the roof. Due to the introduction of the glazed ladder trusses plenty of light flows into the building. For the sometimes problematic points where the masts pass through the roof, this provides an elegant solution. At the ends the glazed walls were joined with the roof, so that the membrane there acts as a canopy.

Plan view, elevations and perspective view of the roof 1:500: 1 main membrane, central bays, 2 edge bays, 3 membrane roof over side blocks, 4 elliptical ladder trusses, 5 main mast, 6 tension ties, 7 edge mast with guys, 8 lens-shaped, glazed opening over side blocks

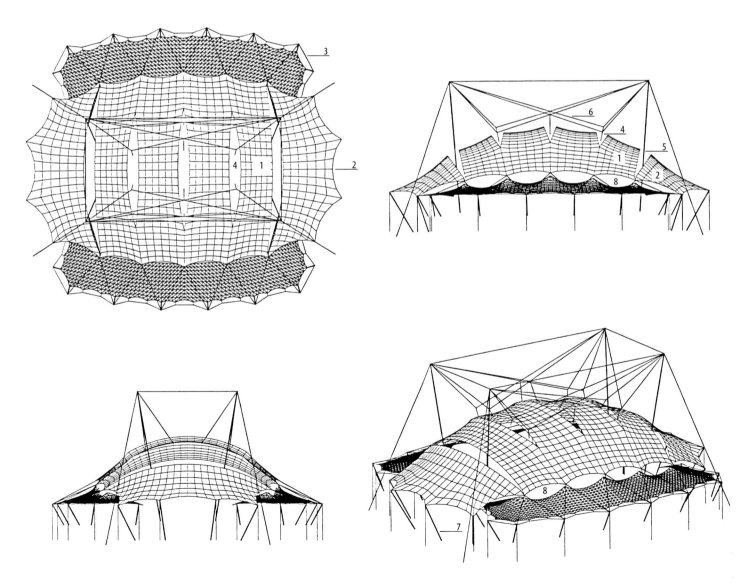

Membrane edge, corner detail and substructure over the lens-shaped, glazed openings

Structure

Four elliptical ladder trusses (chords CHS Ø168.3 x 5 mm with reinforcement tubes Ø 88.9 x 5 mm, bracing/web members Ø 88.9 x 5 mm and tension diagonals Ø 19 mm) are suspended by high-strength tie members (Ø 30 mm) from four masts (Ø 457 x 40 mm). The edges of the roof are held at the canopies by raking struts and over the side blocks by raking A-trestle frames (Ø 193.7 x 10 mm). The tips of the A-frames are joined with the ladder truss by steel cables; the lower, thermally insulated ceiling structures of the side blocks are suspended from the A-frames. Between the A-frames glazed lens-shaped trusses are arranged at the eye-shaped openings, which form the upper edge of the side roofs. The outer edge of the side roofs is held by props and raking struts from within and is tied down by vertical tension anchors to tension piles in the ground.

The structural behaviour of the roof and its tensile structure is complex, especially through the interaction of two different structural systems: the main roof over the sports area and the roofs of the side blocks. Variable external loads result in elongations of the hangers, and due to their special geometry this would lead to unacceptably large displacement at the top of the A-trestle frames. To improve their behaviour prestressed high-strength tension ties (Ø 19 mm) were introduced, joining the internal compression members with the side blocks, so that load variations would change the prestressing in this lower tension tie system. This new suspension system is stiffer than the original support structure and thus limits the movements. The tension ties are mostly high-strength prestress tendons Macalloy 460 (Ø 19 mm to Ø 95 mm), the steel cables are spiral strands (Ø 13 to Ø 22 mm).

The tie foundations consist of tension piles.

Membrane | The roof membrane is of PTFE-coated glass fibre fabric. At the edge cables (spiral strands 1 x 19, Ø 21 mm) the membrane is connected by means of aluminium clamping strips and intermittent metal straps (at 400 mm c/c). The cables under the membrane, joining the tips of the ladder trusses with the membrane corners (7 x 7, Ø 25 mm) and with the edge cables, are connected with the A-trestle frame and the substructure by a welded corner plate assembly.

Interior view with structure of the side blocks

Part plan of main roof 1:200: 1-8 same as on page 40, 9 A-frame, 10 tension tie, 11 main-edge cable (connected to it and not shown are the garland edge cables of the membrane roof of the side blocks), 12 substructure of the steel side-block roof (suspended ceiling)

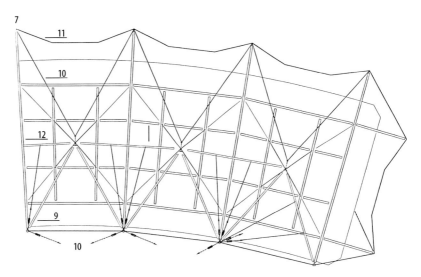

Part plan of the side block at roof level 1:200: 1-8 same as on page 40, 9 A-frame, 10 tension tie, 11 main-edge cable

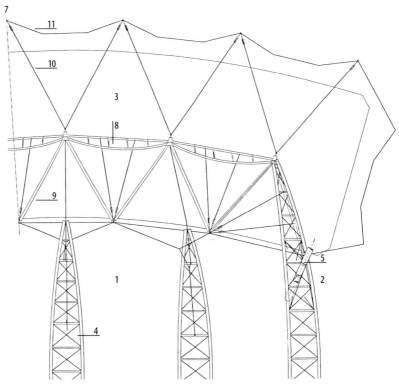

View

Research Laboratory Venafro, Pozzilli, Italy

Design

Requirements | For over thirty years the M & G group has been working in the field of polymer research and polymer processing. In 1990 they decided to concentrate the different research activities at one location, at the enterprise M & G Ricerche SpS in Pozzilli in southern Italy, and at the same time to streamline their activities.

The centre consists of two areas: a technical area with pilot plants for the development of production and processing methods, and a chemical-physical area with labs for the synthesis and analysis of chemical products. The implementation of these chemical and physical large-scale experiments has a varying space requirement which is predictable only with difficulty. From this fact resulted the requirement to create an open, column-free space as large as possible, which at the same time would allow for separate tests to be carried out in smaller protected units.

Concept | Considerations with regard to the optimal shape of the hall led to an elliptical plan; a lightweight, tent-like form appeared already on the first sketches and developed to an oval shape (85 x 32 m) as one large volume, covered by a lightweight structure with a height of 15 m, supported by symmetrical lattice arches and braced by six longitudinal stabilising cables.

The outline of the service installations and the pilot plants led to the choice of the arch shape as a flat basket (three-centre) arch. The maximum height of the pilot plants and the apex height of the building compliment each other. The optimal span for the roof membrane lies between 12 m and 15 m and this determined the number of arches. Aspects like the bracing of the arches, the transfer of the membrane forces and the search for a balanced design caused the architects to arrange the arch planes so that they intersect at a point which at the same time is the centre of the circle connecting the apexes of the arches.

Functions

This space lit by the translucence of the roof membrane and through the glazed edge openings houses research areas. The spaces at the gable ends between the side membrane and research spaces serve as common rooms and reception areas for visitors. The floor space totals 2700 m^2.

Building site – situation | The building is situated like an island in the middle of a rectangular pool and follows the outline of the foundations of a convent which once occupied this site; of the

Client
SINCO Engineering s.p.a.

Architects
General and membrane structure:
Samyn et Associés S.P.R.L., Brussels
(A. Charon, M. D. Ramos, P. Samyn,
M. van Raemdonck, B. Vleurick)

Structural engineering
Membrane and structure: Büro IPL-
Ingenieurplanung Leichtbau GmbH,
H. Mühlberger, U. Rutsche
Substructure: SETESCO S.A.,
J. Schiffmann

Specialist consultants
HVAC: F. Invernizzi
Project management: Studio H
(Arturo Cermelli)

Contractors
General Contractor: SICOS
Subcontractors membrane and steel:
Canobbio spa

Completion
1992

Interior view; pilot plant and office building; membrane roof with truss arch and stabilising cables

original buildings on the site a chapel and a dry-stone masonry aqueduct still exist. The artificial lake serves as a fire well and cools and animates the environment through natural evaporation and reflections. The lake, lined all around with old olive trees, is bridged over at the longitudinal sides of the building by two access ways, which form the main entrances to the building.

Structure

Arches | Six arches carry the roof membrane; they are three-chord lattice arches in the shape of a basket (three-centre) arch. Their triangular cross section varies over the arch length, with a maximum size at the apex, and tapers toward the arch support. They consist of 1764 single tubes in 441 different lengths and configurations. The arches are joined by six prestressed cables under the membrane. These stabilising cables are connected to the arches through pyramid-shaped outriggers to stay clear of the membrane curvature.

Substructure, foundations | The continuous reinforced concrete ground slab with foundation pads

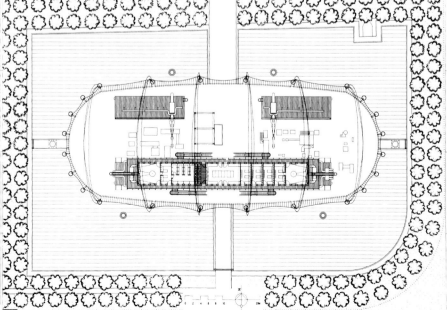

Plan with office buildings 1:1000

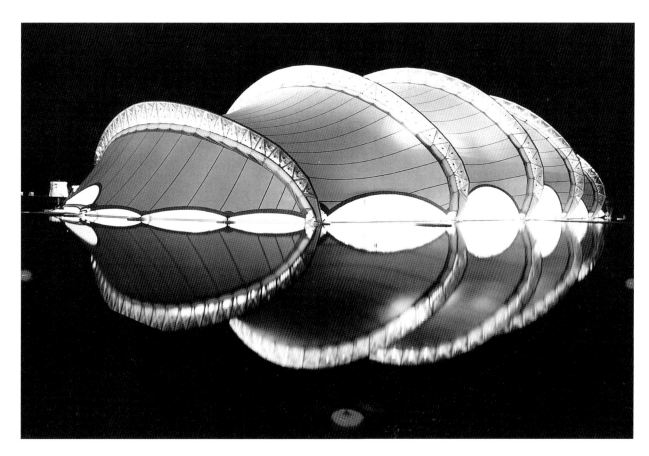

Night view with pool

and strips under the walls of the lab buildings and under the arches were designed for a live load of 20 kN/m². The arch feet are supported on the water level plain on reinforced concrete foundations, which are separated structurally and visually from the ground slab.

Membrane | The membrane material of the roof is a PVC-coated polyester fabric with a tensile strength of 150 kN/m in warp and weft (fill) direction, which is equivalent to a type-4 membrane in the German classification.

The membrane is prestressed between the arches and the edge cables connecting the arch feet. The edge cables are run in membrane sleeves reinforced by webbing. At the arch supports they are connected with an adjustable connection made from a perforated flat steel plate. Along the arches the membrane is connected adjustably through a garland cable with top and bottom cornerplates and threaded U-bolts. For waterproofing a membrane apron is connected to the roof membrane on the outside by a zipper along the edge cable and is clamped to the steel arch along the bottom chord tube. The arches are covered by a transparent, prestressed plastic sheet, thus weather-protecting them economically and simultaneously keeping the arches visible and transparent. The PVC-sheet is stabilised in the triangular bays formed by the arch diagonals by a system of steel disks and prestressed ties.

Facades

Behind the edge cables, there are planar, glazed and arched facade elements. The connection between the roof membrane and the half-arches of the facade consists of a flexible PVC membrane apron, fixed between the upper edge of the facade and the edge cable, which can be tensioned by a horizontal lacing. In the end bays of the membrane the space is closed by a transparent membrane tensioned between roof membrane and floor slab.

Services

Lighting | The truss arches are covered with a clear transparent PVC-membrane, providing additional natural lighting from above and keeping the arch structure visible at the same time. Cover strips and aprons are also made from transparent PVC-sheet. Due to the translucent

Longitudinal section 1:1000

Cross section 1:1000

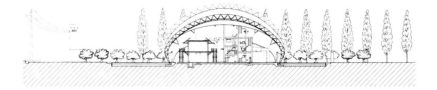

Detail sections at building edge 1:75:
1 truss arch, steel tube, 2 pyramid-shaped steel tube outrigger for stabilising cable, 3 stabilising cable, 4 roof membrane, 5 transparent PVC sheet covering stabilised with steel disks and ties, 6 free-standing facade element, 7 horizontal facade apron, 8 membrane edge with edge cable, 9 reinforced concrete arch support and foundation, 10 lab buildings, 11 installation ducts, 12 ventilation duct, 13 pool basin, 14 reinforced concrete slab with strip foundation, 15 floor ducts

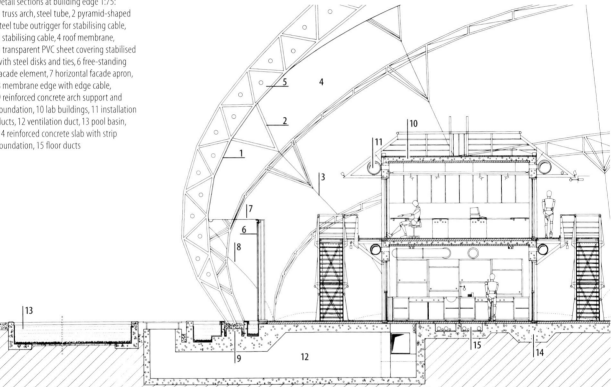

membrane work is possible by day without artificial lighting. At night the building is lit by indirect halogen lighting against the roof membrane radiating from the balustrades of the lab buildings.

Ventilation, air-conditioning | For functional and safety reasons, the large air volume under the membrane roof must be airtight. By means of a simple ventilation system (on one longitudinal side the supply air comes in and on the other side the exhaust air is extracted), through shading by the roof membrane and due to the lake, an air-conditioning was not required despite the high outside temperatures. The inside temperatures correspond to ones under a free standing, naturally ventilated tent, i.e. the outside air temperature in the shade. Offices and lab spaces have independent air-conditioning

Interior view; access bridge to office building; membrane roof with truss arch and stabilising cables, glass facade braced with encastréed I-sections

Bird's-eye view

Recreational Clinic, Masserberg, Germany

Design

Shortly after German reunification the architect was asked to develop a concept for a recreational centre in Masserberg in Thuringia with a recreational clinic and a medical treatment area. The basic idea of the design is the conservation of two typical GDR 'bed houses' (sleeping accomodation) and their connection by a new building which was to be two floors higher. On the east side, in front of the bed houses and the new building lies the leisure complex covered by a membrane roof. The wedge-shaped "pieces of glass" of the new, fully glazed building push into the membrane roof and join the adjacent buildings.

The leisure complex with a plan area of approx. 3200 m^2 extends over the entire length of the new reception building and contains leisure functions and therapy and catering areas. The theatre and the physiotherapy building are partly covered by the membrane roof; their flat roofs serve as terraces. According to the architectural concept this area should differ in style and character from the purely functional bed houses and the fully glazed entrance building and become a symbolic expression of what recreation and leisure mean. So the idea was born to cover this area with a large membrane structure, with a vertical glass facade as a space- enclosing element.

Structure

Masserberg is located at approx. 800 m above sea level in an exposed position with high wind speeds and snow loads. The design snow load of 2.65 kN/m^2 is rather severe, thus six internal masts were needed instead of the originally planned four. The membrane is a self-supporting tensile structure, a high point membrane with inside masts, edge cables and guyed edge masts. In the bays between the high points it is tied down by groove cables. The inside masts are joined through security cables with the edge guys to secure the masts in case of catastrophic failure of the roof membrane.

Steel structure | The primary structure consists of 6 main masts, 33 edge supports and 9 edge ties around the re-entrant "piece of glass" building. The main and edge masts are supported on different levels and thus have greatly differing lengths. The main masts consist of a steel tube (Ø 508 x 20 mm and Ø 508 x 14.2 mm) with mast lengths of 8.4 m to 17.2 m. At the base the masts are supported by spherical steel bearings, set in a spherical cut-out in the lower part of the bearing which is cast into the floor slab. At the high points the six main masts have two

Client
Rennsteig Kur- und Touristik GmbH

Architects
General: Architekt Laurens Schneider-Zimmerhackl
Membrane structure:
IPL-Ingenieurplanung Leichtbau;
Project architect
DI Architekt Michael Kiefer

Structural engineering
Membrane and primary structure:
IPL-Ingenieurplanung Leichtbau
Steel structure: Ingenieurbüro Stendtke + Vedder
Facade: Büro Gerritz

Specialist consultants
HVAC: Ingenieurbüro Eberhardt Möller + Partner
Building physics: Büro Paul Schröder

Contractors
General contractor:
ARGE Hochtief-Oevermann-Groh
Sub-contractor Membrane:
Canobbio spa
Steel – primary structure:
Stahlbau Zwickau
Steel – facade:
Stahlbau Heinrich Weller

Completion
1993-94

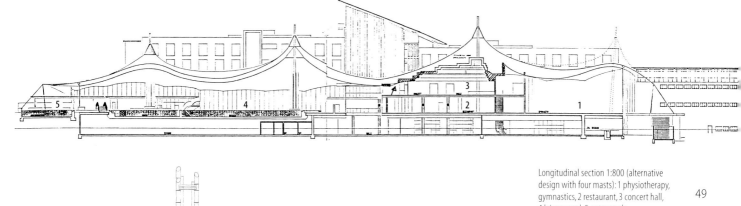

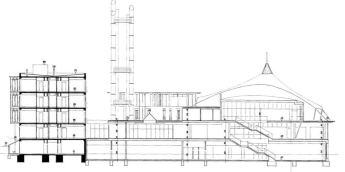

Longitudinal section 1:800 (alternative design with four masts): 1 physiotherapy, gymnastics, 2 restaurant, 3 concert hall, 4 leisure pool, 5 outer pools

Cross sections 1:800

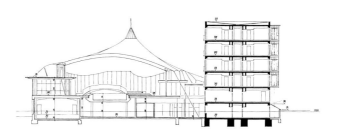

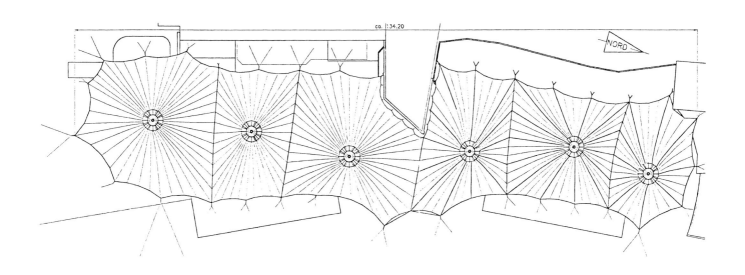

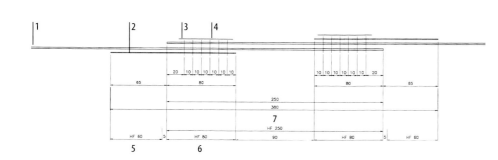

Roof plan 1:800 (final design with 6 masts)

Combination seam for PVC-coated polyester fabric type VII: 1 PVC-coated polyester fabric type VII, 2 PVC-coated polyester fabric type III, 3 PVC-coated polyester fabric type I, 4 sewn seam (6 x, at 10 mm c/c), 5 HF-seam, width 60 mm (HF = high frequency), 6 HF-seam width 80 mm, 7 HF-seam width 250 mm

Model photograph: Alternative design with four masts

East view

membrane support rings each, made from steel tube (outside ring Ø 3.82 m, inner one Ø 2.26 m). The outside ring carries the outer membrane. To carry the membrane forces a circular steel plate ring is welded onto the CHS for bolting on the membrane clamping strip. Due to the extremely high membrane forces at this point the clamping strip is additionally held by external steel straps. The membrane ring is supported via 12 hanger cables from a prestressing suspension above on the main mast. It can be stressed from a set of anchor plates on the mast head through threaded bars. The lower anchor plate serves as fixing for the threaded bars; the upper plate is jacked up by a prestressing jack, thus introducing the required prestress. Similarly the lightweight, internal membrane is fastened to the inside ring and fixed by a clamping ring. The lower ring is welded via steel plate spokes to a CHS sleeve. Both rings are closed with a sheet metal diaphragm. The diaphragm of the upper ring is shaped to enable natural ventilation of the membrane space through louvres and additional forced ventilation through extractor fans if necessary. The sheet metal floor of the lower ring has a built-in access hatch, which can be reached via an access ladder along the mast. The upper ring is covered with a conical sheet metal hat of 16 folded sheet metal elements joined by a welt. It includes an external access door for maintenance of the fans and for a possible restressing of the membrane. The tip is formed by a steel tube with largely decorative function.

The edge supports of the external membrane are formed by different structures:

By tripod structures with guys, connected either to single pad foundations with tension anchors or by anchor parts cast into the reinforced concrete floors, and at the glass building via steel cables, connected directly to reinforced concrete corbels.

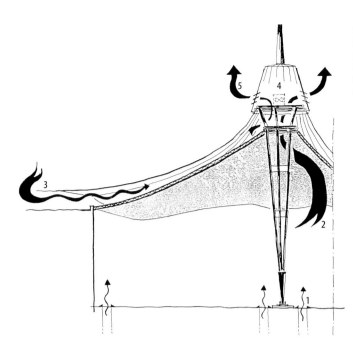

Detail section: Ventilation of the membrane roof (design variation with lattice mast): 1 supply air, 2 exhaust air from interior, 3 Ventilation of membrane space and thermal insulation, 4 fans, 5 exhaust air

The tripod guys come in different versions:
– one compression strut, two tension ties,
– two compression struts, one tension tie,
– three struts capable of resisting tension and compression.

A typical edge support consists of a steel tube pinned at the foot. For the guy cables restressing has been made possible by appropriate detailing of the mast heads. The edge supports follow the course of the membrane edge and are either on the building, connected to building parts and/or on separate foundations or connected to injection anchors. They are situated on different levels, on reinforced concrete floors or pad footings.

Foundations, tension anchors | For anchoring the guy cables and for transferring the column load steel inserts were provided and cast into the reinforced concrete. Guys forces of edge supports on pad footings are carried into injection anchors (micropiles).

Double layer membrane with thermal insulation | The roof has a double-layer membrane; the outer membrane extends over the entire area (3300 m^2), the internal membrane covers the leisure pool (1270 m^2) and the sports area (395 m^2). It follows the outline of the outer membrane. The internal membrane supports an insulation layer; the air space in between is ventilated. A minimum distance of 50 cm between the membranes had to be maintained under load, to allow the air circulation of min. 5000 m^3/h. The built-up of the double layer membrane is:
– Outer membrane = waterproofing layer
– Aerated cavity
– PE-cover sheet
– 16 cm mineral wool as double layer insulation, laid with staggered joints
– Inner membrane

Inner membrane | The outer membrane consists of 15 panels, joined during the erection stage with clamp strips. The membrane material of the outer skin is a new PVC-coated polyester fabric (Diolen Superfest) specially developed by Verseidag for high-strength applications in wide-span air halls with 60 000 to 120 000 m^2 floor area. It has a tensile strength of 20 kN/50 mm and 19 kN/50 mm (warp/weft). For comparison: The standard membrane material for wide-span membrane structures, type II (to German specification), has a strength of 3.9 kN/50 mm and 3.5 kN/50 mm, the rarely used type V has 9 kN/50 mm and 8 kN/50 mm. The internal membrane is a PVC-coated polyester fabric (Diolen) type III; together with facade and building walls it forms a largely airtight space enclosure. It is connected to the smaller internal high point ring by a clamping strip. At the edge the inner membrane follows the facade, it is connected there stressed over a steel pipe, forming the upper edge of the facade structure.

Membrane apron | To keep birds and other small animals from entering into the membrane cavity, it had to be closed by a membrane apron along the edge of the inner membrane. This apron must follow the displacements and movements of the different load cases and at the same time allow air circulation under the roof. The displacement under changing loads relative to the prestressed state amounts to calculated 1.5 m; for safety the maximum distance between inner and outer membrane at the edge was determined at 2.0 m. In the interior the clear space is smaller, being determined by the required minimum ventilation cross section and by the minimum clearance for installing the thermal insulation (0.7 m).

The upper part of the membrane apron consists of a porous grid fabric, its length corresponds with the minimum distance required to accommodate the movements. The additional length is taken up in a concertina arrangement of fabric in the bottom part, so its length may be extended up to the maximum value of 2.0 m. The concertina folds are held in place by diagonally tensioned elastics.

Assembly

The assembly took place in late fall and early winter. To allow erection of the membrane the substructure shell had to be completed in the area of the membrane roof, as well as the facade construction with the cladding largely in place. Access for a mobile crane was only possible from one side, from the east. Edge and inner masts were erected first, the inner masts were placed over the previously positioned high point rings. To facilitate erection and prestressing of the membrane the edge masts were inclined towards the inside. Scaffolding was erected for the assembly of the 15 membrane panels which for transport reasons are approximately equal in size. The membrane was lifted into place in one day. It was prestressed by moving the edge supports out using tirfors and by jacking the upper mast rings up. For erection of the inner membrane it had to be laid out roughly at the level of the facade's upper edge requiring scaffolding in some areas. The inner membrane was prestressed by jacking up the high point rings and from the edge masts.

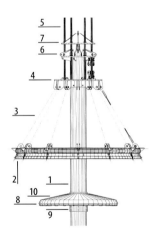

High point: 1 main mast (Ø 508 x 20 mm or Ø 508 x 14.2 mm), 2 outer ring (Ø 3.82 m), 3 hanger cable (12 of), 4 prestressing suspension, 5 threaded bars, 6 anchor plate, 7 jacking plate, 8 inner ring (Ø 2.26 m), 9 tube sleeve, 10 sheet metal covering

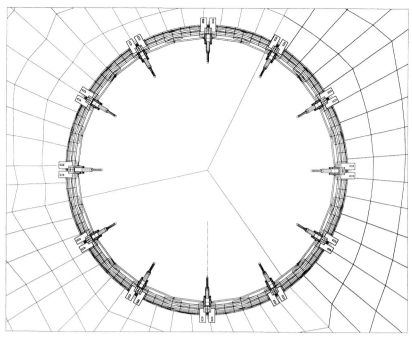

Outer high point ring, 1:50

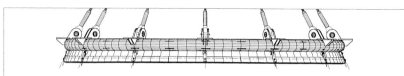

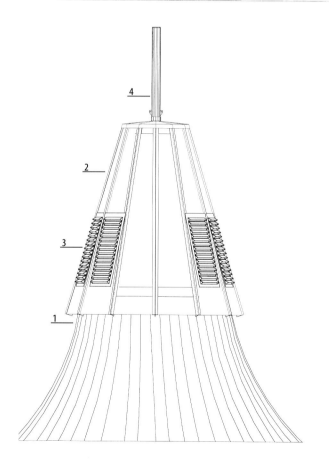

Sheet metal high point cover (tin hat) 1:100: 1 outer membrane, 2 folded sheet metal elements joined by a welt, 3 ventilation grill, 4 steel tube mast tip; not shown: outer access door

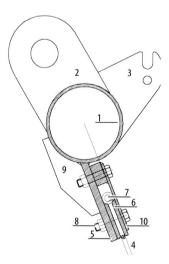

Membrane connection at the outer high point ring 1:8: 1 outer ring (steel tube), 2 connection plate for hanger cables, 3 connection plate for sheet metal covering (tin hat), 4 outer membrane (PVC-coated polyester fabric type VII), 5 steel plate ring, 6 clamping strip, 7 boltrope, 8 fixing screws, 9 stiffener, 10 outer connecting plate

Butterfly house; roofscape with ventilation duct

View of entrance with round entrance lobby

Client
Prof. Dr med. Klaus-Werner Wenzel

Architects
General and membrane structure:
Horst Dürr, IF Ingenieurgemeinschaft Flächentragwerke, Constance

Structural engineering
Membrane, structure and substructure: IF Ingenieurgemeinschaft Flächentragwerke, Constance
Membrane: Flontex GmbH

Completion
1993

Butterfly House, Berlin, Germany

For the federal garden exhibition BUGA 1982 in the Britz district of Berlin a "church in the green" was built as a membrane structure. In 1993 it was converted for semi- commercial use as butterfly house and extended by several rooms. After 1982 the church remained unused for a long time and was ultimately destroyed through vandalism. In the years 1991 to 1992 a physician and butterfly lover searched for a location for a butterfly house and through the park and garden administration had this building recommended to him. Because of its dilapidation and the weathering of the membrane and due to the special climatic requirements of a butterfly house only the old primary structure was used and a new, thermally insulated membrane roof was planned by the same architect who designed the original BUGA "church in the green".

Brief

The brief contained:
- a central area for the large flight room with visitor's paths and planting;
- an entrance lobby with cloakroom, ticket office and cash register;
- a large showroom with displays of books, pictures and replicas of exotic butterflies – but no pinned butterflies (!); set in the room divider separating it from the large flight room are tubular perspex (acrylic glass) butterfly incubators where visitors can observe hatching;
- various ancillary rooms as well as
- an outside heating and climate container with a gas tank next to it.

Design

The refurbishment was in fact virtually a new construction but keeping two of three of the original glue-lam arches and the rear foundations. The existing oval plan shape of the hypar roof was added to by various extensions so that in plan it resembled a formal combination of the wings, body and head of a butterfly. Refurbishment and new construction were carried out allowing for the particular climatic requirements for raising and keeping exotic butterflies using modern building materials. Special importance was given to the room climate in the flight room with a required inside temperature of 26 °C and a relative humidity of 86%. In addition a roof skin permeable for UV radiation was required to create a healthy, pest-free climate for plants and animals. This was achieved through special roof and wall components: a clear double layer UV-permeable roof skin in the sunlit area, a thermally insulated, opaque membrane on the shaded north-west roof, and a hard cladding of transparent and thermally insulating polycarbonate twin web panels in the curved steep wall area under the outer glue-lam arch, all fulfilling the U-value requirements. The rear wall cladding forms a warped, transparent and thermally insulating polycarbonate surface.

Old 'church in the green' at the federal garden exhibition BUGA

View and plan 1:200 with round entrance lobby, entrance area, showroom, store and ancillary rooms (office, preparation, personnell), large flight room, technology containers and gas tank

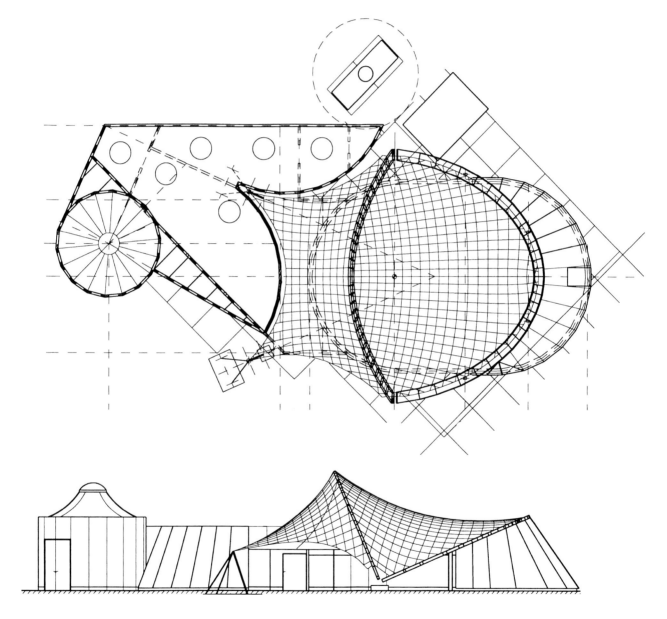

Interior view with roof and transparent wall

Structure

Two arches with differing cant form the main structure. Above it in the main area a double-layer membrane is stressed and is guyed towards the front entrance by an edge cable and support struts. The horizontal arch is supported by four vertical columns made from rectangular hollow sections. The polycarbonate glazing is fastened to sloping tubular mullions, which are pin-supported at the foundation and are fixed to the arch above through a movable tube sleeve, so that no vertical loads are carried into the foundations. The ancillary rooms have a wood frame structure with a flat roof. The structure of the entrance lobby consists of a wall in wood frame construction supporting a double-layer membrane roof and a suspended flying strut supporting the high point.

Membrane | The outer roof of the butterfly flight room consists of an open weave glass fibre fabric with a mesh width of 5 mm, laminated on both sides with a 100-μm thick FEP sheet. The inside roof is a plain, unreinforced ET sheet with a thickness of 120 μm, high translucency and a UV-permeability of 80%. The smaller roof consists of PVC-coated polyester fabric, which was laminated with PVC foam for better thermal insulation. The edge of the outside roof exterior is formed by a movable tube frame to safely apply the prestressing forces, which are relatively high for this type of fabric. The connection to the arch is formed by a membrane apron. The membrane strip is stretched by hand during installation and fastened under tension. This is a critical method which is not always recommended. It was chosen because the jacking distance was very short. Its success depends on the skill of the assembler; a re-tensioning is impossible.

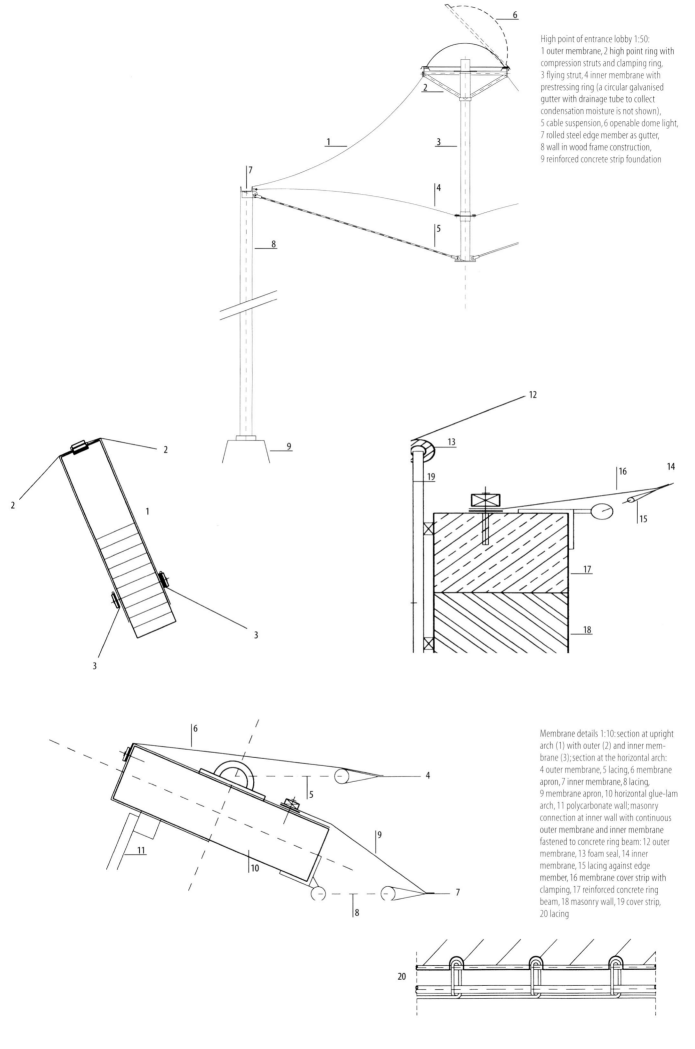

High point of entrance lobby 1:50:
1 outer membrane, 2 high point ring with compression struts and clamping ring, 3 flying strut, 4 inner membrane with prestressing ring (a circular galvanised gutter with drainage tube to collect condensation moisture is not shown), 5 cable suspension, 6 openable dome light, 7 rolled steel edge member as gutter, 8 wall in wood frame construction, 9 reinforced concrete strip foundation

Membrane details 1:10: section at upright arch (1) with outer (2) and inner membrane (3); section at the horizontal arch: 4 outer membrane, 5 lacing, 6 membrane apron, 7 inner membrane, 8 lacing, 9 membrane apron, 10 horizontal glue-lam arch, 11 polycarbonate wall; masonry connection at inner wall with continuous outer membrane and inner membrane fastened to concrete ring beam: 12 outer membrane, 13 foam seal, 14 inner membrane, 15 lacing against edge member, 16 membrane cover strip with clamping, 17 reinforced concrete ring beam, 18 masonry wall, 19 cover strip, 20 lacing

Interior view; stair hall with portal frames and suspended stair landing

Extension of Lycée la Bruyère, Versailles, France

Design

Which architectural concept, which style would be suitable for a scientific secondary school, situated only a few steps away from the Royal Palace at Versailles and bearing the name of a famous social critic, author, scientist and 17th-century member of the Académie? For the architect the 'equation' to solve contained not only these unknowns but also the problem of integrating the old school buildings erected between 1823 and 1938.

The architect Zbinden made up his mind: His decision for modernity, in its widest sense, for him was the only culturally acceptable answer, since historic re-creation carries the inherent danger of doing badly that which our ancestors did well. To create quality with criteria and possibilities which the past did not possess, is for him an obligation which we must fulfil without nostalgia.

And then there is still the park, a natural element of the existing ensemble, which creates a connection between the architectural styles covering one-and-a-half centuries. The new building with its terraces takes up the rhythm of the topography, the distinctive sloping site, and culminates in a pitched roof unusual for this rural architecture.

Through its terraced stepping down of the floors a skylight illuminating the depth of the building becomes possible, which allows a development of the classrooms in transverse direction instead of longitudinally, a layout which brings the pupils in closer contact with the teacher and his teaching aids.

The main building materials are steel, glass, and wood: simple materials, designed with much care, creating a cheerful atmosphere with a measure of luxury, which is valued by the pupils and tolerated by the authorities.

The atmosphere is determined by different spatial elements:
– by the PTFE-coated glass fibre fabric covering the stair hall,
– the underground dining room, sunk into the terrain,
– the great hall, and
– the class rooms opening towards the park.

Situation | The new building is situated on a south slope above the old building, behind which an underground dining room was added. A wide outside staircase joins the old and new build-

---

Client
Région Île de France

Architects
General and membrane structure: Walter Zbinden Atelier d'Architecture et d' Ingeniérie, Versailles and Friedeman Kugel, Konstanz (tensile structure)

Structural engineering
General: Starrs
Membrane, structure and substructure: Büro IPL-Ingenieurplanung Leichtbau and Stromeyer Ingenieurbau, Konstanz

Contractors
General contractor: Hervé, Entreprise Général
Subcontractor membrane: Stromeyer Ingenieurbau, Konstanz

Completion
1995/7

Front view

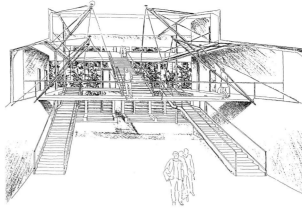

Perspective view; stair hall; shown is a structural alternative for stair suspension and membrane roof which was not executed

Section perpendicular to slope 1:200

Ground floor plan (part) 1:500

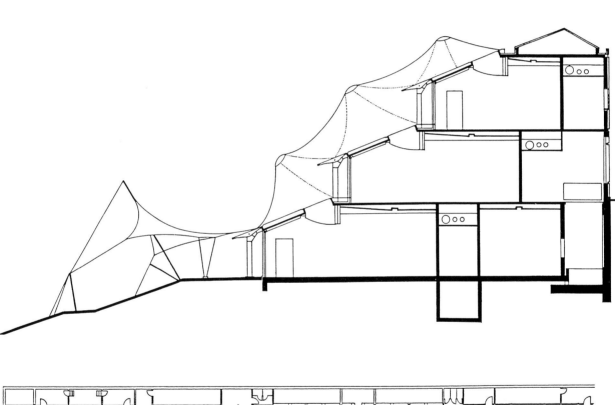

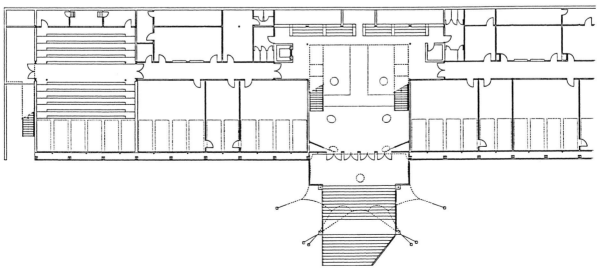

Interior view: stair hall

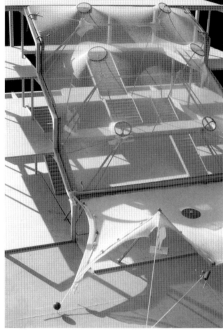

Model photograph

ings and leads directly up into the wide stair hall, connecting to the corridors and class rooms on its sides. The building has three storeys; the upper floors are set back by 4.6 m following the slope. On the 1st upper floor the building has access from the street above.

Structure

The structure of the building is a galvanised and painted steel frame of purlins, I-sections and CHS columns. Frame members, beams and columns, are HEB 260, the inner columns are CHS Ø 168.3 x 4.5 mm and Ø 219.1 x 5.6 mm. The structural bays on both sides of the stair hall have diagonal bracing.

The high points of the membrane roof are carried by short compression members arranged roughly perpendicular to the membrane. They transfer their load into vertical portal frames with inclined steel tube columns, guyed in front by round bar ties with turnbuckles. The columns on the first floor also support the loads from the bridge-shaped stair landing. They carry their load diagonally in the respective floor structure.

Membrane and substructure | The membrane consists of two parts, the main membrane over the stair hall and a roof canopy over the entrance; it is made of PTFE-coated glass fibre fabric, type B 18089 (Verseidag Indutex), with a tensile strength of 5.0/4.5 kN/5 cm. The main membrane follows the shape of the building's front edge. Six high points in the membrane surface, with high point ring and dome light covering, form a humped membrane surface. Along the building edges the membrane is connected to the primary steel structure via steel I-sections, running as gutters along the front edge above the facade. At this edge the main membrane is prestressed into the steel gutter by a tensioning device, and similarly the front canopy. At the front the canopy is supported by a free cable edge on two masts with guys and tied down in the middle by three cables and a steel ring (Ø approx. 1.0 m) forming a low point in the membrane, thus creating the necessary surface curvature. At the high points the roof membrane is fastened to a steel high point ring with bolts and clamping strip. An inclined strut (CHS Ø 101 x 6.3 mm) with a built-in top and bottom universal joint carries the high point ring (Ø 1.0 m), adjusting itself independently perpendicular to the membrane plane. The strut at its lower end is inserted into a tube sleeve, welded to the steel portal frame.

At the membrane corners of the canopy two triangular steel plates connect the fittings of the edge and guy cables with each other.

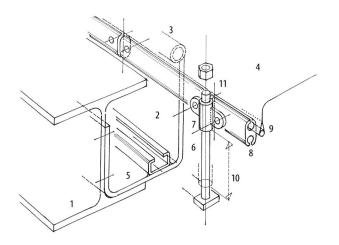

Detail section: Prestressing gutter: 1 structural frame, steel I-section, 2 continuous cold worked gutter, 3 welded-on steel tube as edge protection for the membrane, 4 roof membrane, 5 Halfen rail, 6 square-headed bolt M20, 7 sleeve strap as flexible connection for 8 boltrope section, aluminium-extrusion, 9 membrane edge with welded-in boltrope, 10 prestressing lift 150 mm, 11 flexible joint in boltrope section

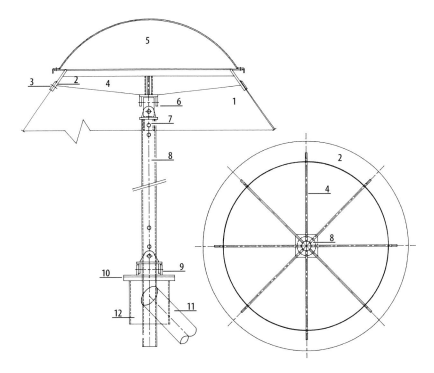

Section and plan of high point 1:20: 1 roof membrane, 2 conical sheet metal edge, 3 clamping ring, 4 ring spokes, 5 polycarbonate dome light, blown, 6 upper universal joint, bolted, 7 rotation joint about strut axis, secured with bolt, 8 compression strut (CHS Ø 101 x 6.3 mm), 9 lower universal joint, bolted, 10 flange connection (at the same time separating two trades), 11 primary structure, tube frames, 12 node, steel tube with plate stiffener (stiffener not shown for clarity)

Only some of the guy cables of edge masts and membrane corners proceed directly to the foundations. The intermediate membrane corners are both anchored on a tripod of one mast and two guys, to avoid a 'trip trap' resulting from very long and low cables, with the additional advantage of removing it from potential vandals.

Facade

The conventional curtain wall facade is set back relative to the steel structure; a tube member parallel to the facade carries a translucent, adjustable sun screen.

Overall view with lions' open-air enclosure

Carnivore and Palm House, Munich Zoo, Germany

Design

A novel zoo concept: In the 'jungle tent' visitors as well as small apes and birds can move around freely in a natural setting. Separated only by glass panes from the visitors, the animals can be observed as if they were in a forest glade. A little stream with waterfall, walk-on 'hills' and sumptuous tropical and subtropical vegetation create the impression of a visit to the jungle. For the natural growth of the plants and their protection from parasites and diseases maximum levels of UV-light are necessary; and the minimum temperature may not drop below 12 °C. These considerations soon led to investigations of wide-span structures and transparent cladding. In several design steps the solution of a cable-net roof emerged with a cladding of pneumatic foil cushions. An alternative cladding system exploring overlapping small polycarbonate panels (of the kind used on the concert sail in Radolfzell) had to be excluded because of its large air-permeability and the resulting unsatisfactory thermal insulation.

Situation, plan | The construction site is situated in Munich Zoo at the foot of the Harlaching hill in a landscape protection zone, in direct proximity of the historically important and listed elephant house of 1911. The building has an octagonal plan defined by curved edge cables; it is symmetrical about one axis. Its covered plan area is 1140 m^2. Two visitors´ entrances furnished with air locks are situated on the north-west side. The animal boxes for lions, panthers, jaguars and other large carnivores are arranged along the facades, some of them having access through the facade into open-air enclosures. On the longitudinal side facing the entrances lies an artificial planted hill, under which a number of bird boxes and other ancillary rooms are located together with the delivery entrance. In the middle part, on both sides of the mast foundations, there is a plant area without underlying ground slab.

Structure

Two 17-m high masts carry the prestressed net structure made from high-grade steel cables with a mesh width of 2 x 2 m. It is supported on two cable-braced tubular steel masts; a straight compression tube is stabilised through a three-chord tie bracing of round steel bars with star-shaped spreaders made from short compression struts arranged along the mast length. At the mast head the ridge and eye cables are connected, which at the same time laterally support the mast tip. At the base the masts are pin-supported on the reinforced concrete substructure. The edge masts consist of simple tubular steel columns.

Client
Münchener Tierpark Hellabrunn AG

Architects
General: Büro Herbert Kochta Architekt BDA, project architect E. Lehner
Landscape architect: Büro Teutsch + Partner
Cable net structure: Büro IPL-Ingenieurplanung Leichtbau

Structural engineering
Cable net and membrane: Büro IPL-Ingenieurplanung Leichtbau (design); KOIT High-Tex GmbH (general contractor); Schlaich, Bergermann + Partner (in cooperation with the contractor)
Reinforced concrete structure and substructure: Ingenieurbüro Dieter Herrschmann

Specialist consultants
HVAC: Ingenieurbüro Bergbauer
Thermal/moisture insulation: Prof. Schaupp, Grünwald
Term planning, site supervision: Ingenieurbüro Zinner & Sohn
Installations: Ingenieurbüro Bergbauer
Electrics planning: Ingenieurbüro IBE Rolf Günther

Contractors
General contractor & roof: KOIT High-Tex GmbH
Facade: Glasbau Seele

Completion
1995

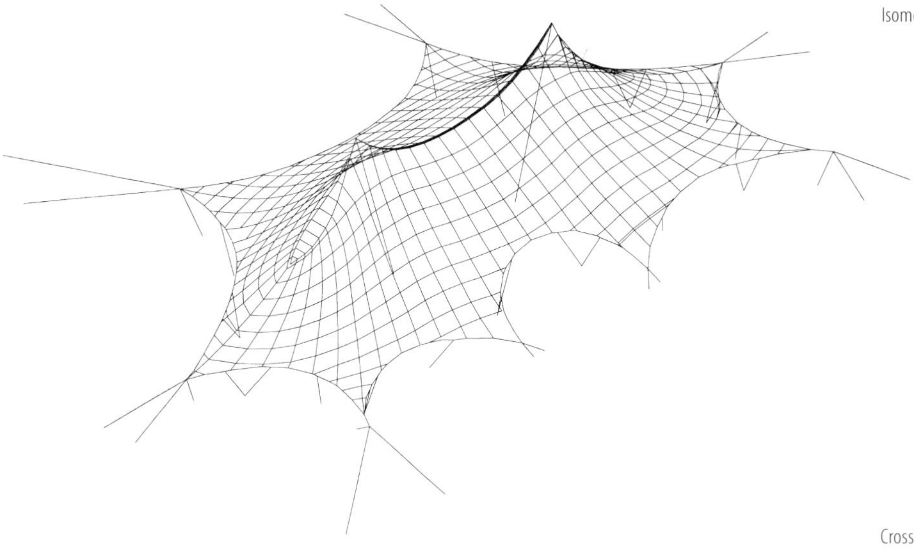

Isometry; cable-net roof

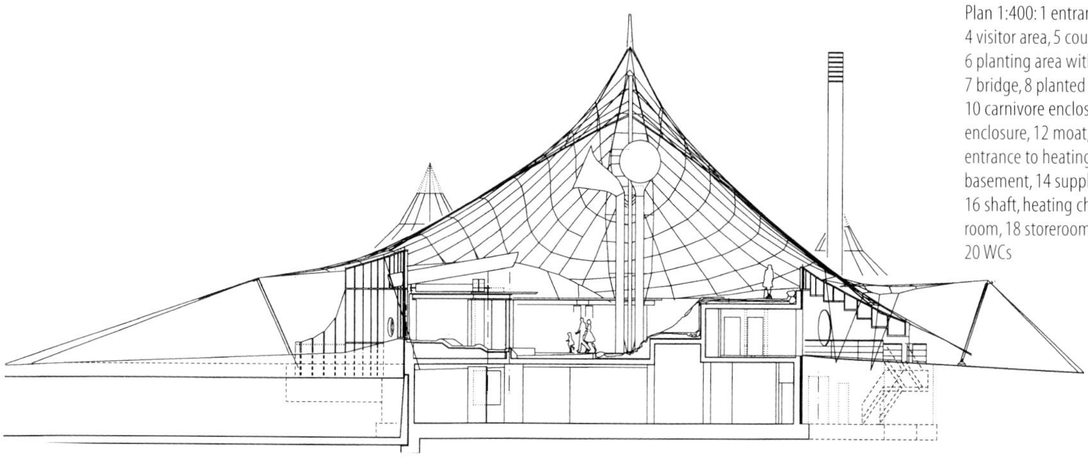

Cross section 1:400

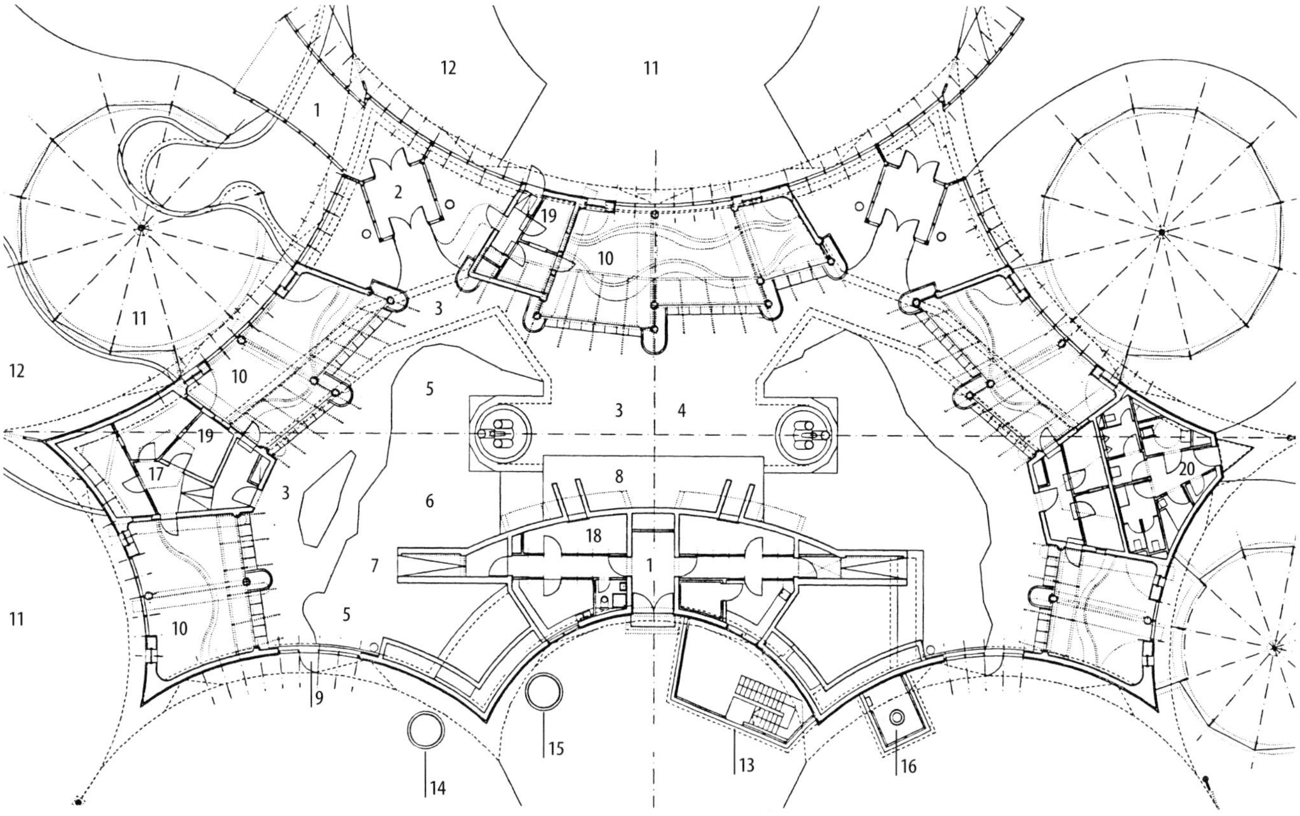

Plan 1:400: 1 entrance, 2 air lock, 3 path, 4 visitor area, 5 course of stream, 6 planting area without ground slab, 7 bridge, 8 planted hill, 9 emergency exit, 10 carnivore enclosures, 11 open-air enclosure, 12 moat, 13 supply shaft and entrance to heating and technology basement, 14 supply air, 15 exhaust air, 16 shaft, heating chimney, 17 attendant room, 18 storeroom, 19 cubbing room, 20 WCs

West view with visitor entrance

Part view with lions' open-air enclosure

Cable net | The roof is a symmetrical cable net surface with two inner high points (H = 17 m) joined by ridge cables. The structure is a radial cable net made from radial and hoop cables. It is composed of double radial cables (Ø 18 mm) and single hoop cables (Ø 16 mm); at the top it is supported by four ridge cables (Ø 32 mm), held at the edge by edge cables (Ø 36 mm) and anchored at its eight corner points by masts, guy cables (Ø 36 mm) and injection anchors. Snow loads, if they are not melted already by the warm air, are carried by the internal pressure into the lower foil of the pneumatic cushions and into the radial cables of the net, which introduces its load mainly into the ridge cables connecting the masts. The statutory snow loads to DIN 1055 were fully allowed for in the structural calculations. Wind loads are carried by ring and radial cables into edge cables and substructure. The net cables are joined through movable high-grade steel cable clamps, onto which the edge clamping arrangement of the cushions are fastened. The cushions have been developed into cutting patterns along edge and surface seams and are connected via a boltrope edge and by special aluminium-edge profiles to the cross clamps of the cable net and to the edge cable clamps. The cable clamps of radial and ring cables are connected with a central bolt and are free to rotate, so that the mesh angles can adjust freely to adapt to the three-dimensionally curved cable-net surface.

The net edges are 3.5 m above the ground; there the net cables are equipped with a swaged closed cable fitting which is pin-joined through a flexible one-bolt connection with two forked connecting elements, which in turn are fastened to a two-piece, circular edge cable clamp. At the net corners the edge cables are connected via fork fittings directly to the head plates of the edge masts.

The four ridge cables running from one mast head to the other are joined at the junctions with the radial net cables through a welded cable clamp. The net cables are connected to its lower plate stiffener by a swaged fork fitting and a flexible one-bolt connection. Under the ridge cable and suspended from it run a cable tray and the supply pipes for the pneumatic cushions. The ridge is closed on top through a grating, serving as a service walkway, and by a double-layer of ETFE-foil above it connected with the membrane cushions along their edge. This unorthodox solution allows walking on the ridge and at the same time provides a transparent and thermally insulated ridge covering. The ridge cables are continued on the other side of the mast down to edge guys, forming eye shaped openings, which are closed with mechanically operated ventilation louvres.

Foil cushions

The foil cushions are composed of two fluoropolymer foils (t = 0.2 mm). The foil material, ETFE (Ethylene Tetra Fluor Ethylene) is anti-adhesive, highly self-cleaning and is classified as 'difficult to ignite' (class B1 to DIN 4102). Its durability has been shown through long term tests, and, according to company information, shows no (considerable) changes of optical or mechanical properties after 10 years of natural weathering. The translucency of the roof structure in the waveband important for plant growth totals approx. 90%. The translucency of the foil itself in the UV range amounts to 96%. The 64 wedge-shaped cushions (max. b = 2.5 m, max. l = 18 m) are filled individually with dried and purified outside air via separate, flexible and closable cushion feeds through a pressure pipe running along under the ridge, which helps to prevent condensation and to preserve the translucency of the cushions. The inflation pressure varies according to the external loading, wind and snow, and is controlled by a computer system, which can compensate minor damage of the membrane through a pressure increase. The pressure normally amounts to 250 pa and can be increased to 350 pa if required. For the inflation of the cushions altogether 4 fan units are provided, with one main and one reserve blower. The latter serve to ensure the safe working of the system even in case of breakdown of a main blower, and also serve to increase the air volume in case of damage to the membrane. The air supply is controlled centrally using an anemometer located at one of the masts.

The aluminium edge sections which hold the transparent foil cushions at the edge proceed from the ridge radically down to the lower net edge. These High-Tex-aluminium extrusions (consisting of three principal aluminium extrusions and an extruded cover strip section) are

Opposite page:
Interior view with ETFE cushion, facade sealing tube, net eye and animal enclosures; clearly visible are the welds of the clear transparent foil cushions

bolted directly onto the radial cables. The ring cables are connected with a spacer piece to the radial cables underneath and are located under the foil cushions, so that these can deform unhindered under load. The 'belly cables' common in earlier cushion structures were not used here. (They are, however, used along the edge, to hold the sealing tube in place).

Facade connection | The wind and rain-proof connection of the foil cladding with the heated steel and glass facade must be able to tolerate large deformations of the roof, e.g. under snow load. The connection is made through curved tube cushions, which are fastened to the upper edge of the facade, held in position with cables and pressed against the roof cushions by their inflation pressure. In addition to the air supply controlling the pressure, relief valves are installed, which open at a sudden pressure increase under load and thus allow a fast pressure adjustment. A further internal sealing tube, running parallel to the edge belly cables provides an almost perfect air seal and at the same time prevents the escape of small tropical birds.

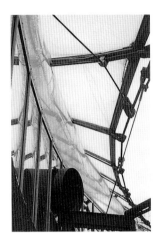

Ventilation | On two sides of the masts ventilation ducts proceed to the air intake funnels located at the mast heads. In the winter months these air intake funnels serve as forced ventilation. As the free-flying tropical birds in the jungle tent are very draft-sensitive, the air speed had to be kept much lower than in similar ventilation systems. Through the funnel shape the air cross section is expanded which decreases the air speed. According to the HVAC calculations a 10.5-fold air exchange per hour is sufficient for the single cell cushion roof, so that even given extreme outside temperatures the inside temperature does not move outside the required tolerance band. During the summer months the building is ventilated by natural stack effect through two large net openings. They consist of lens-shaped areas cut into the membrane and closed by acrylic glass louvres individually supported by stainless steel brackets sealed to each other. They can be moved continuously and have their aperture angles regulated electronically. The system can be opened so that the entire net eye area is available as a ventilation opening.

Assembly

For the installation of the roof structure a scaffolding was erected in the interior, from which most of the work could be accomplished. After positioning of the masts the cable net was assembled on the scaffolding from single cables connected at the nodes by cable clamps and raised gradually. The net was prestressed from ridge and edge cables to 8 anchor points. The prefabricated cushions were fixed first along one side of the High-Tex cushion profile, rolled out sideways, tensioned and finally fixed on the other cushion side. The edge areas inaccessible from the scaffold had to be worked from the net. The entire construction period totalled 3 years.

Edge mast with net connection, cushion profile and sheet metal gutter

Facade connection with ET-sealing tube

Detail view: Facade corner with ET-sealing tube and belly and auxiliary cables

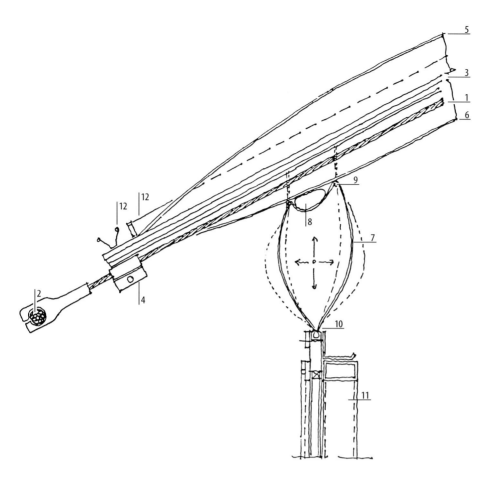

Facade connection with ET-sealing tube 1:10: 1 net cable (twin radial cables Ø 18 mm) with fork fitting and edge-cable clamp, 2 edge cable (Ø 36 mm), 3 ring cable (Ø 16 mm) with 4 cable clamp, 5 foil cushion upper membrane, 6 foil cushion lower membrane, 7 facade connection; outer sealing tube, 8 facade connection; internal sealing tube, 9 belly cables for securing the sealing tube (the tube cushions are laced on at the belly cables), 10 facade connection section, 11 metal and glass facade, heated, 12 bird protection

Ridge connection 1:10: 1 ridge cable (4 x Ø 32 mm), 2 ridge-cable clamp, 3 bolts, 4 radial cables (2 x Ø 18 mm), 5 fork fitting, 6 High-Tex cushion section, 7 air-filled foil cushion, 8 cable tray, 9 air supply, 10 flexible cushion air-supply tube, 11 walk-on ridge, grating, 12 ETFE-foil cover, 13 safety hooks, 14 bird-defence cable, 15 clamping plates for transparent ridge covering, 16 service bracket

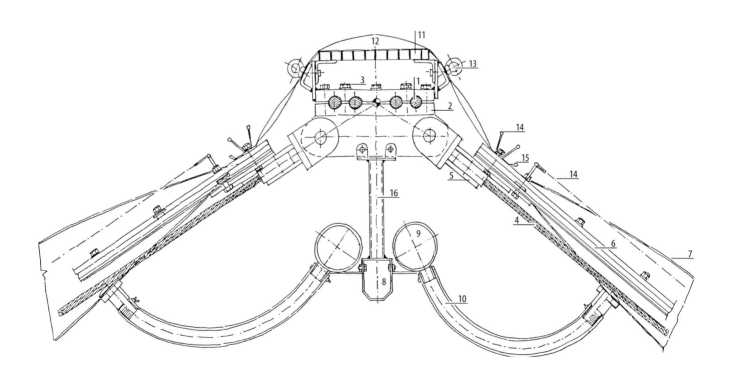

Diagonal view

High-Rise Facade, Osaka, Japan

Design

Clouds and water, symbolically represented as aluminium and fabric panels in a spatially curved curtain wall facade of a high-rise building: The Japanese architect Keizo Sataka wants to visually and formally create an expression of softness and tenderness, which contrasts with the angular shapes, hard materials and sharp edges of modern cities. Three-dimensional curved surfaces are intended to give the city a new, soft skin. As an expression of transition the architecture combines the standard aluminium facade panels of today with innovative panels of coated fabrics. Whether his architectural philosophy stands up to close scrutiny is one matter; the built architecture, however, is fresh and convincing.

Plan, use | The building has a more or less rectangular plan with 9 storeys, 3 basements and a 4-storey penthouse. The leisure complex includes restaurants, bars, showrooms and a disco with a total floor area of 1978 m^2.

Structure

The high-rise structure consists of a three-storey reinforced concrete basement and composite steel and concrete decks in the upper floors. The design wind speed (typhoon load) is 60 m/sec (216 km/h). A wind tunnel test was not considered necessary.

Membrane panels and substructure | The aluminium and fabric panels are rain screen panels of a curtain wall facade in front of a space-enclosing and thermally insulating inner facade of polycarbonate panels. They are equipped with a bracing steel frame. All panels were prefabricated in the works, where the fabric membrane was prestressed against the steel frames. It is one of the first applications of coated fabric as a curtain wall facade. (Exhibition Hall 4 in Friedrichshafen, Germany is another, earlier example; it was erected around 1970). For Japan it is also the first building in which PTFE-coated glass fibre fabric was permitted in a fire protection zone.

The membrane elements consist of PTFE-coated glass fibre fabric (Shearfill II from Nitto Chemfab Co., Ltd.). As the facade was developed on a triangular grid the prefabricated panels consist of a changing number of grid triangles. They are braced by a welded steel frame (U 150 x 75 x 6.5/10 mm, CHS Ø 48.6 x 2.3 mm). At the edge the membrane is turned around the steel edge tubes of the frame and anchored in the steel U-section by a clamping strip (Fl 50 x 12 mm) and a welded on boltrope. At the clamps and turning points it is padded by neoprene

Client
Morique Building Co.

Architects
General and membrane structure:
Keizo Sataka + Institute of International Environment
Structure: Sun Corporation

Structural Engineering
Buildings: Taisei Corp.
Membrane: Institute of International Environment and Taiyo Kogyo Corp.

Contractor
Buildings: Joint venture Taisei Corp + Simon Construction Co., Ltd.
Membrane: Taiyo Kogyo Corp.

Completion
1992/93

Front view

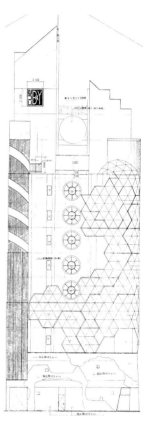

View 1:500

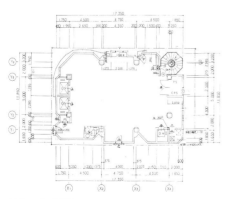

6th floor plan 1:500

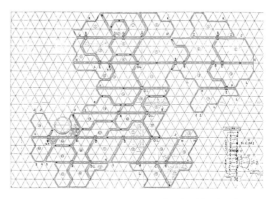

Developed facade grid 1:500

Detail section 1:10; membrane connection and panel frame: 1 membrane, 2 steel edge section (U 150 x 75 x 6.5/10mm), 3 CHS (Ø 48.6 x 2.3 mm), 4 neoprene padding strip, 5 boltrope, 6 outer neoprene weather strip, 7 cold rolled U-section (U 40 x 40 x 3 mm) with neoprene gasket (zipper), 8 hexagonal connection plate (steel plate t = 12 mm), 9 floor edge member (U 200 x 80 x 7.5/11 mm), 10 reinforced concrete floor (t = 150 mm), 11 steel web (t = 12 mm), 12 polycarbonate panel, 13 clamping strip (fl 50 x 12 mm), 14 connection plate (t = 9 mm)

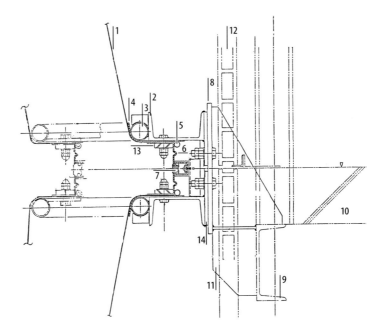

Detail view 1:10; membrane connection and panel frame: 1 membrane panel, 2 steel-edge member (U 150 x 75 x 6.5/10 mm), 3 connection plate (t = 9 mm), 4 hexagonal connection plate (steel plate t = 12 mm), 5 steel web (t = 12 mm), 6 bolt M16

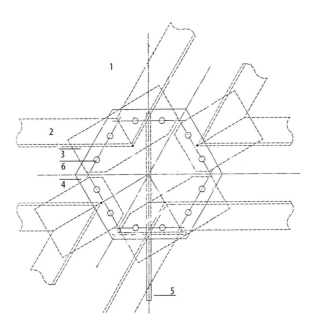

strips, which also provide a wind and weather-tight seal. The panels are sealed and connected to each other by means of outer weather strips running up to a central cold rolled U-section (U 40 x 40 x 3 mm), where they are closed with a neoprene zipper gasket. The panel frames are welded to a hexagonal plate connected via a steel plate (t = 12 mm) to the floor edge member (U 200 x 80 x 7.5/11 mm) and are thus anchored in the reinforced concrete floor behind (t = 150 mm).

View of the marina

Dubai Creek Marina Club, Dubai

The Yacht Club is part of a larger building complex, which contains both a golf club and a golf academy and a chandlery with a boat workshop. It lies directly on the waterfront above the Dubai Creek, which here is already tidal. The contract for the planning of the golf club was awarded as the result of an international architectural competition, while the design for the Yacht Club was developed in direct contact and co-operation with the client. Although originally to be designed like the golf club, the client later decided to aim for a completely different, modern architectural concept, which ultimately led to the existing design.

Design

Since the golf club symbolically represents an Arab dhow, the traditional two-masted boat of the Gulf region, the architect concluded that in analogy the Yacht Club should resemble a modern ship. The building is composed of three parts, each shaped like a ship´s prow; the parts are of unequal size and have a common internal service core. A continuous glass facade along the entire building is made possible by the clover-leaf plan of the building. The facade is curved following the building´s perimeter and leans out at 15° to the vertical, so that the ship-shaped character of the building is reinforced. The roof terrace is covered by two overlapping sails made from coated fabric which provide shade to the area below. They are carried by six outside masts made from steel tube, which are completely separated from the building. The tube masts are also inclined outwards, at the same angle as the glass facade, and are guyed toward the outside by steel wire ropes and anchored in concrete foundations, some of which are under water. Each floor has a 'promenade deck', whereby the marine theme is continued in the interior. The design of the steel cable structure and its anchorages posed particular problems, since the guy ropes, which become very hot in the summer heat, should be arranged in such a way that nobody can touch them nor run into them by mistake; they should, nevertheless, run close by the building so that they could be read as part of the structure and the marine theme. From the desired position of the building directly above the water resulted the necessity of an effective corrosion protection in this hot and humid marine environment. On the waterfront, directly in front of the building and above the water, extending over the entire length of the building, there is a seating deck made in wood planking, supported on piles in the Dubai Creek.

Plan, functions | The upper level of the building houses a restaurant with bar and a large acrylic

Client
HH Sheikh Mohammed bin Rashid al Maktoum, Crown Prince of Dubai

Architects
General and membrane structure: Godwin Austen Johnson; Chartered architects, Stourport-on-Severn, Great Britain, and Dubai

Architect and project partners: Brian Johnson; RIA project architect: Adrienne Simpson

Structural engineering
Membrane: Ingenieurbüro für Tragwerksplanung Dr. Ing. Hans-Joachim Schock, Constance, , and Stromeyer & Wagner GmbH, Constance

Structure: Ingenieurbüro für Tragwerksplanung Dr. Ing. Hans-Joachim Schock, Constance, , Stromeyer & Wagner GmbH, Constance and Robert Matthew Johnson Marshall, Dubai

Substructure: Robert Matthew Johnson Marshall, Dubai

Specialist consultants
HVAC: Robert Matthew Johnson Marshall, Dubai

Quantity Surveyor: Dove Associates, Quantity Surveyors

Contractor
General contractor: Al Naboodah Laing, Dubai

Membrane and steel: Stromeyer & Wagner GmbH

Facade and glazing: Alico, Sharjah

Civil engineering and underwater works: Al Naboodah Contracting, Dubai

Pontoons: Simmoneau, France

Completion
1993

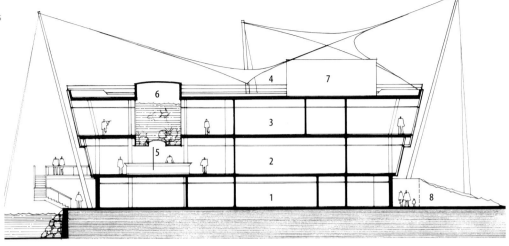

Section 1:400: 1 ground floor, locker rooms and toilets for club members, services, personnell, storerooms, 2 1st floor, shop, adjoining rooms, administration, 3 2nd floor, restaurant + bar, 4 roof terrace with bar, 5 bar, 6 fish tank, 7 bar, sale room, 8 entrance/circulation

fish tank, 5.0 m in diameter. The entrance floor contains administration rooms and a bar, with a view up into the fish tank directly above through its dome-shaped acrylic floor. The basement contains service rooms, kitchens, staff rooms and locker rooms for the club members.

Construction

The building was erected using the fast track method, as were the other buildings of this complex. The design was approved and released for execution in November 1991; 14 months later, in January 1993, the Yacht Club was opened.

Structure

The roof structure is a self-supporting, tensile membrane. Membrane forces are carried by edge cables which run in membrane sleeves; corner plates and short tension members connect them to the masts. The guy cables of the masts proceed from the connection points of the sails to foundations in the ground and to concrete-gravity anchors under water. Two of the masts have four guy cables to reduce the bending moments resulting from the second sail and its forces, connected in the middle of the mast length. This results in a complex, three-dimensional mast structure. The outer corner masts C and E and the masts on the building's axis of symmetry, masts A and D, have two guy cables and carry one sail each. The two masts B and F on opposite

2nd floor plan, 1:400: 1 restaurant, 2 fish tank, 3 dining area, 4 cocktail bar, 5 guest staircase, 6 personnel staircase, 7, 8 WC, 9 kitchen, 10 bar, 11 bar storeroom, 12 ventilation /service duct, 13 electrical cabinet, 14 elevator

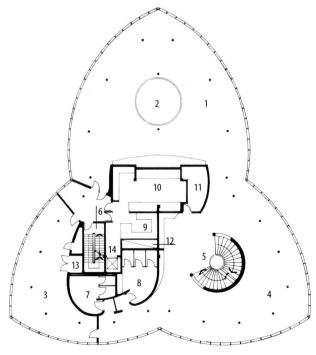

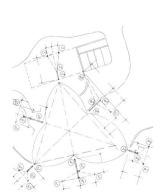

Key plan of masts and guys

Sails on the roof terrace

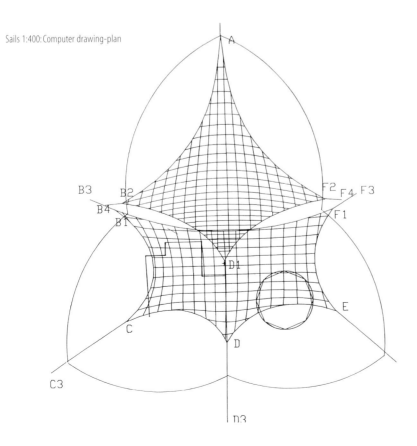

Sails 1:400: Computer drawing-plan

longitudinal sides carry two sails on different levels, so that these masts are guyed by four cables. The welded tubular steel masts (Ø 273 x 5.0 mm to Ø 609 x 8.8 mm) were fabricated in one piece of galvanised structural steel St52 with mast lengths of 10.3 m to 22.0 m, with a welded mast head and gusset plates to receive the cable fittings. At the mast base the welded bearing assembly was equipped with an installation sleeve to simplify the introduction of the mast bearing into its bottom counterpart. The design windspeed is 47 m/sec (169 km/h).

Tension members | All tension members, i.e. edge cables, guy cables and membrane ties are stainless steel spiral strands (material grade 1.4401), which in parts were executed as twin cables for reasons of availability. Stainless steel cables, unlike galvanised cables, are not always available in all listed dimensions at short notice, often they are only available with long delivery times or must be specially made. A way out of this dilemma is to use available cable sizes (usually smaller cables are more easily available than larger ones) in a twin arrangement and with sufficient strength, of course. The guy cables and membrane ties are connected to the masts through welded steel plates and fork fittings. The lower ends of the mast guys are connected adjustably to twin stainless steel threaded bars (joined by steel spacer plates and embedded in the reinforced concrete foundations), by means of cable end terminals and stainless steel crossheads made from square stainless steel bars. They are connected to the welded steel plate assembly of the mast head via swaged stainless steel fork terminals.

Foundations | Some of the tension foundations are below sea level in this tidal area of the Dubai Creek. At these underwater anchorages the threaded bars are long enough, so that the connection between cable and threaded bar is above the highest tide, and thus can never be washed over by the sea water. The underwater gravity anchors on the sea bottom are dug-in precast concrete elements; they consist of a concrete base plate with shaft rings above, which were filled with underwater concrete. In the area of the seating terrace the guy cables pass through a deck opening protected with a railing.

Membrane | The roof membrane consists of two overlapping sails, one with 5 and one with 4 corners, both of PTFE-coated glass fibre fabric. For technical and architectural reasons the sails are very shallow, resulting in fairly large membrane forces despite the relatively small membrane size. The edge cables are carried in membrane sleeves. They are connected with nuts and threaded stainless steel fittings to the membrane corner plates; the cable force is introduced through a short steel tube welded onto the corner plate. The membrane corners are composed of a lower triangular steel plate with short rolled steel brackets welded on to receive the U-bolt, a clamping plate above and the membrane sandwiched in between and fixed in position with a stainless steel boltrope (or rather a bolt 'bar') inserted in a membrane pocket. The U-bolts are twin threaded bars with a steel plate welded in between.

View with cable detail

View with prestressing device at the arch end and 'wishbone' guy structure

Ministerial Leisure Centre, Riyadh, Saudi Arabia

Design

The swinging forms of the double-layer membrane roof and the glass walls of the leisure centre contrast pleasantly with the rectangular geometry of the Ministry for Municipal and Rural Affairs, on which it was erected as a penthouse structure. It creates a relaxed, luxurious atmosphere. With its shape the roof reminds one also of the traditional Bedouin tents of the Arabian peninsula, which form an essential part of the local cultural heritage.

Situation, use | The penthouse structure houses a number of different leisure areas for the employees of the ministry, ranging from a common room, to various sports areas and a sauna. The large free span allows a flexible space concept. The different areas form enclosed spaces in an open-plan area. For the interior design and fittings floor and wall tiles made of natural stone from local quarries were used, together with cherry-wood furniture custom-made in the USA. A special requirement was that the building should not be seen from the ground. Its maximum inside height is 6.7 m with a covered plan area of 716 m^2.

Air conditioning | A ventilation floor serves as plenum for the air conditioning, with air outlet grilles along the external walls. In the squash courts additional air is blown in from above. An inner skin of glass fibre fabric improves the effect of the air conditioning and thus helps to save energy.

Structure

Six planar arches form the primary structure which carries the double-layer membrane roof with a maximum span of 18.2 m. The arches in plan are arranged perpendicularly to the curved longitudinal axis of the building and are supported by bifurcating tree columns made from steel tube. Two smaller, symmetrical arches with two simple tree columns are arranged at the narrow building sides; four larger arches, arranged in pairs in the middle part of the building, have one of their supports on the roof slab and the other on a single-storey blockwork construction, which contains changing and ancillary rooms. Statically the arches are two-pin frames; the arch supports are pinned perpendicularly to their plane which meant they could be assembled lying on the floor and then set up vertically.

Prestressing devices for the membrane are connected at the arch ends, to enable later tensioning along the arches.

Client
Kingdom of Saudi Arabia, Ministry of Municipal and Rural Affairs

Architects
General, membrane structure and interior design: FTL/Happold

Structural Engineering
Membrane and structure: FTL/Happold

Specialist consultants
HVAC: Buro Happold
Lighting design: Thomas Thompson Lighting Design

Contractor
General contractor: CERCON
Membrane: Birdair
Steel: Tubemasters
Cables and fittings: Pfeifer Seil- und Hebetechnik GmbH & Co
Facade: Vistawall Architectural Products

Completion
1994

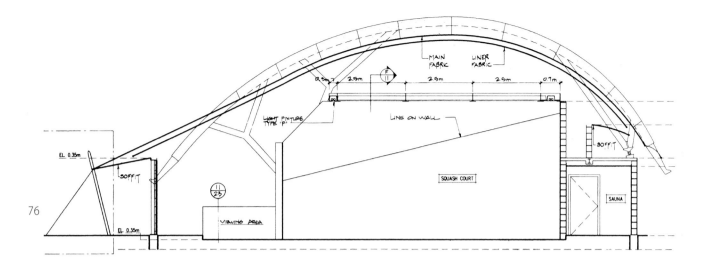

Section 1:150
Plan 1:500

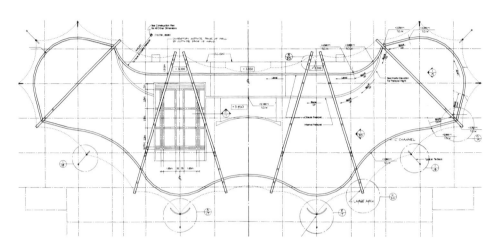

Guy structure | Between the arches the membrane is either tied down through a simple cable guy structure with a short steel tube mast or else through a 'wishbone' structure. This is composed of two curved steel tube columns, which together form a pointed arch (= wishbone).

Roof membrane | The internal and external roof membranes are made of PTFE-coated glass fibre fabric; a PVC-coated polyamide fabric forms the connection between facade and outside membrane.

Production, Assembly

During the design phase a model was built out of cork slabs, wire, stretch fabric and flexible plastic bars for the arches. After approval from the Ministry the final design was carried out in cooperation with several offices in the USA and in Europe. The glass fibre fabric was made in New York state, the steel work came from England, cables and fittings from Germany, and the glass facade was manufactured in Texas. After the reinforced concrete slab of the existing building had been strengthened, the leisure centre was assembled and erected within a month.

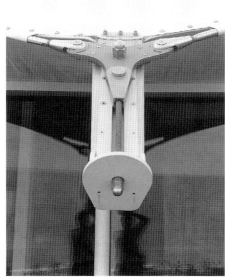

Interior view with tree columns and inner membrane

Prestressing device at the arch end

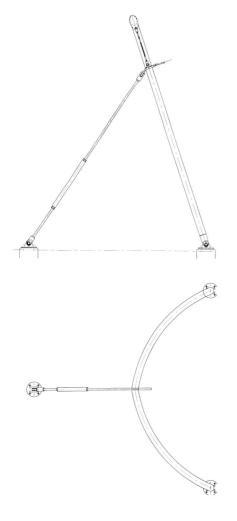
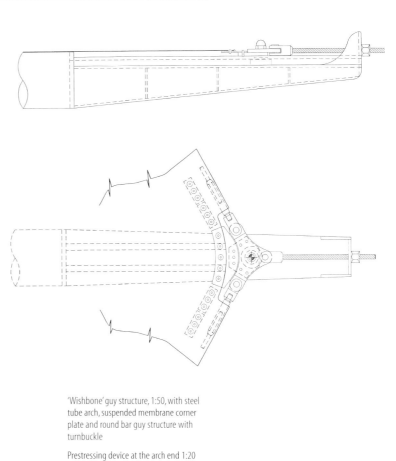

'Wishbone' guy structure, 1:50, with steel tube arch, suspended membrane corner plate and round bar guy structure with turnbuckle

Prestressing device at the arch end 1:20

Covered Tennis Court, Gorle, Italy

Client
Comune di Gorle

Architects
General: Studio Arch. Merlini e Arch. Natalini
Membrane structure: Studio Tensoforma

Structural engineers
Membrane, structure and substructure: Studio Tensoforma

Contractors
Membrane and steel: Tensoforma Trading S.R.L.

Completion
1991

Covered tennis courts and tennis sheds exist in multitudes, but few of them are aesthetically pleasing, and hardly any of them is as beautiful as the one described here.

The design brief for a covered tennis court is clearly defined: The plan dimensions are given by the size of the playing fields together with the required moving space around them, similarly the height is given by the required space clearance in longitudinal and transverse direction. The end walls should be opaque in their lower parts to avoid glare. Natural lighting from above is desirable, and so is transverse ventilation during the hot season.

Design

Two adjacent playing fields are covered each by an independent, self-supporting arch structure, consisting of parallel lattice arches. The arches are supported on steel trestles to achieve the necessary height over the playing field. The trestles along the central spine are connected to form a stiff bracing frame over a central access corridor. The space-enclosing element is a fabric membrane, together with a linear roof light made of curved and planar polycarbonate sheets over the central spine, which cover the upper portion of the central bracing frames. The roof membrane cantilevers along the long sides of the building by way of steel brackets, forming a continuous edge canopy, so that during the summer months the fabric side walls can be removed to aid transverse ventilation without the danger of rainwater entering the building. In the end bays the roof membrane is carried by tubular steel arches which are supported by a concrete wall with a curved upper edge.

Dimensions | The covered plan area is 22 x 36 m for each playing field; the total floor area is 2 x 792 = 1584 m^2. The maximum building height is 8.5 m on the outside; the corresponding inside dimension at the net is 7.80 m. The height of the parabolic top of the end walls varies between 2.25 m at the centre and 3.5 m at the edge.

Structure

The structure consists of a self-supporting roof membrane supported by five two-pin three-chord truss arches (top chord Ø 133 x 4 mm, bottom chords Ø 101.3 x 3.6 mm, diagonal web members Ø 76 x 3.2 mm). Five pairs of trestles at 7.2 m c/c carry the arches at a height of 3.5 m. The compression strut of the trestle has a cantilever bracket attached at its top, to which the edge of the roof membrane is connected. The strut member with a rectangular cross section is

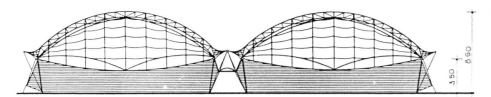

Plan and elevation 1:500

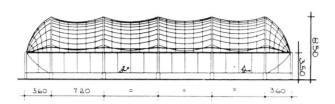

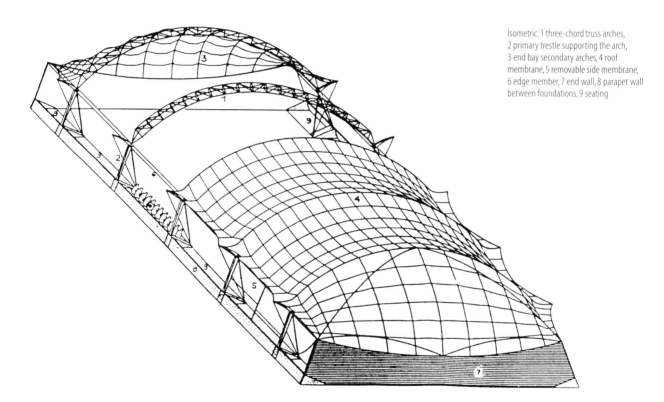

Isometric: 1 three-chord truss arches, 2 primary trestle supporting the arch, 3 end bay secondary arches, 4 roof membrane, 5 removable side membrane, 6 edge member, 7 end wall, 8 parapet wall between foundations, 9 seating

Internal view with three-chord truss arches, three-dimensional cable system for the lateral bracing of the arches and primary and secondary trestles supporting the arches and anchoring the bracing cables

made up of welded steel plate and tapers according to the shape of the moment diagram. The associated tension member is made of steel pipe (CHS) with pin connections at top and bottom.

Between the primary trestles there are secondary CHS trestles (at 3.6 m c/c), to which the valley cables of the cable net, bracing the arches, is connected.

These planar trestles are arranged at a right angle to the side of the building, with a plane dimension of 2.0 m between tension and compression member, and are connected at the top by a horizontal edge member and braced in the end bays by a diagonal pipe bracing member.

The trestles along the central spine form an A-frame made from steel tubes (Ø 88 x 3.2 mm; Ø 76 x 3.2 mm and Ø 60 x 3 mm). For ease of erection it is assembled from five components.

In the end bays the membrane is supported by nine tubular steel arches, supported by the truss arch on one side and connected in pockets at the reinforced concrete end wall.

The roof membrane spans in transverse direction from arch to arch, vertical roof loads are carried into the arches; and the arch thrust is taken into the edge trestles and central A-frames. The external compression members and the inner ties are connected to reinforced concrete foundations. The cantilever brackets support the membrane corners and receive the edge cable forces. The roof was designed for a snow load of 0.6 kN/m^2.

Bracing

The building is braced in longitudinal direction by compression diagonals in the end bays together with a horizontal CHS collector member at the eaves. In transverse direction the building is braced by the arches and the supporting trestles together with reinforced concrete end walls. Perpendicular to their plane the arches are braced by a prestressed, three-dimensional cable system, consisting of five stabilising cables connecting the arches, and a prestressing cable parallel to the arches. This cable system is connected at five points along the arches and to eight secondary edge trestles. In the end bay the cables run across the last arch to be anchored at the top of the concrete end wall.

Membrane | The roof membrane is made of a PVC-coated polyester fabric Precontraint 1002 (Ferrari) with a strip tensile strength of 4.2/4.0 kN / 5 cm (warp/weft) and with spread of flame treatment. It has a translucency of approx. 15 %, which together with the polycarbonate roof light is sufficient for daylight illumination.

Structure during erection with end bay in foreground: end wall, tree-chord trusses and secondary arches carrying the membrane, plus primary and secondary trestles with cantilever bracket to receive the roof membrane

View of structure along the central spine with bracing A-frames during erection; the translucent polycarbonate roof light is not yet in place

Internal view along the central spine with bracing A-frames and translucent polycarbonate roof light

Detail section at spine wall: 1 roof membrane, 2 main arch, 3 bracing A-frame, 4 rc end wall, 5 joist supporting translucent polycarbonate roof light, 6 polycarbonate roof light, 7 sheet metal gutter, 8 planar polycarbonate cladding, 9 sheet metal flashing, 10 ball shield membrane/net apron

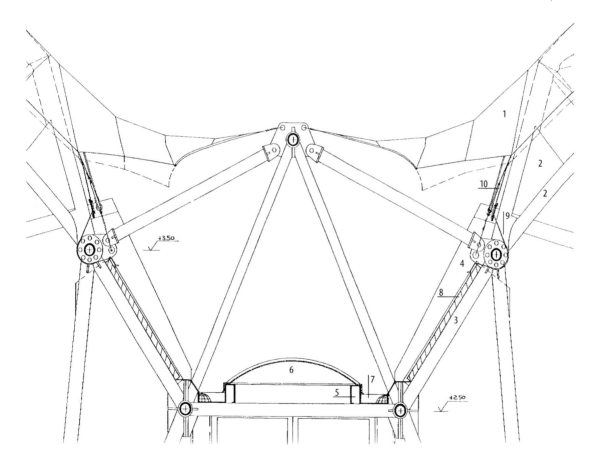

South view at night

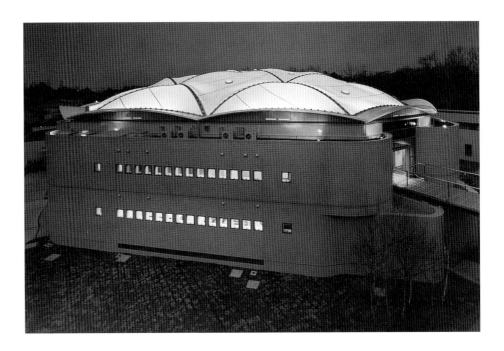

Kaetsu Memorial Gynasium, Tokyo, Japan

Design

The site of this gymnasium for the women's junior college of the Kaetsu University in Tokyo is located in an area classified as residential according to the development plan (class A according to the Japanese building regulations) whereby the building height of residential buildings is limited to 12 m. One of the design requirements of the client, the Kaetsu University, was therefore, that all rules and regulations of the development plan should be fulfilled without waivers and at the same time a usable space with an interior height of 11.5 m as required for national volleyball tournaments had to be created.

In addition the university allowed no excavation into the ground. This left a mere 0.5 m for the roof structure including the roof build up. A further essential aspect in designing a gymnasium is good natural lighting. Also desired was a 'green' building, friendly to the environment, using natural lighting and natural ventilation. Only a membrane structure could fulfil all these demands.

A wide-span, translucent roof illuminates the arena with diffuse, glare-free and still direct sunlight. To match this, the architects wanted to create a visually soft and harmonious space coloured in shades of white. A favourable precondition for this design was that the confidence of planners and client in the possibilities of a membrane roof had recently been strengthened by the successful completion of the Kajima Pneudoms as a translucent roof for a tennis court.

Plan | On a site of 23 300 m² a building emerged with a plan area of 1260 m² and a floor space of 1603 m². The plan shape is square. The entrance areas are situated in the re-entrant and well-rounded corners of the plan square. Above them the plant rooms for ventilation and air conditioning are located. The ancillary rooms are arranged along the sides of the plan square in four narrow, two-storied wings with flat roofs. On the 1st floor there are the locker rooms with showers, toilets and the sauna, the storerooms, training and preparation rooms as well as the rooms for the P. E. instructors; the 2nd floor contains recreational and club rooms.

Structure

The parasol dome is supported by a curved two-way grid structure, rectangular in plan, and made from steel tubes (Ø 139.8 x 6 mm). Frei Otto would probably call it a grid shell. The two-

---

Client
Kaetsu Women's Junior College, Tokyo

Architects
General and membrane structure: Kajima Design

Structural engineering
General and membrane: Kajima Design

Contractors
General contractor: Kajima Design
Subcontractor membrane: Taiyo Kogyo Corp.

Completion
1985/86

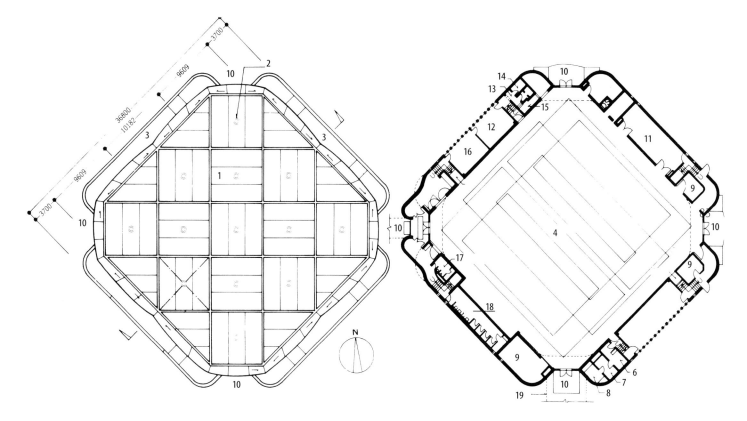

way grid in plan has square and triangular meshes and twin edge arches (Ø 89.1 mm) and is covered with individual panels of PTFE-coated glass fibre fabric. The building is supported on pile foundations. Due to typhoon exposure the design wind speed is as high as 60 m/sec (216 km/h).

Roof membrane | The membrane panels are composed of PTFE-coated glass fibre fabric. They are prestressed by a cable system running diagonally from node to node of the structural grid and fixed to a central membrane ring, so that a sufficient membrane curvature is provided to carry the wind uplift.

Various new structural details and methods were developed, among others
– a detail to prestress the membrane from a fixed edge,
– an aluminium edge clamping section, replacing the traditional membrane clamps with bolts and clamping plates, thus avoiding perforating and weakening the membrane at this point, and
– the design of the central prestressing rings with cables and pulleys.

At the low point the membrane has a circular cut-out with an edge boltrope, which is held in a two-piece ring, made of a circular (annular) boltrope profile. At the lower ring rollers (Ø 36 mm) are connected, pullling down the low point using diagonal cables (spiral strand Ø 10 mm with a 2-mm polyamide sheathing) against the primary structure. The low point rings are closed with an acrylic dome light (Ø 600 mm) above and planar wire glass below.

Along the membrane edge runs a boltrope, anchored in a two-piece boltrope section (b = 65 mm), bolted to two steel angles (L 90 x 7) above and below. The angles are pulled toward the outside by horizontal stainless steel tensioning bolts (below M20, above M12) and against the grid tube via a welded steel plate web with a stiffening steel-flat top chord, thus prestressing the membrane. A neoprene gasket with a zipper profile provides a seal against the neighbouring panel. The seal at the building edge against the GRC shells is achieved in a similar way using a neoprene gasket fixed in a groove and a butyl seal.

Plans, roof and ground floor 1:500:
1 membrane panel with membrane ring, dome light and cable suspension, 2 dome light, 3 flat roof of the reinforced concrete substructure, 4 gymnasium, 5 gymnastics and training room, 6 locker room, 7 shower, 8 sauna, 9 storeroom, 10 entrance, 11 preparation room, 12 instructor room, 13 kitchen, 14 instructors toilet, 15 shower, 16 reception room, 17 men's toilet, 18 ladies toilet, 19 connecting bridge (at 1st floor)

Interior view

Entrance

Aerial view of the university campus with gymnasium

Building construction

For the inside surfaces new types of finish were used which had never previously been used in this form in Japanese gymnasiums, also new materials, among them acoustic wall panels from basswood, finished with a light whitewash coat. Below the acoustic wall cladding there is a new type of basswood insulation material.

Assembly

The scheme design was started in July 1984; the detail design was completed in July 1985; the construction period totalled 8 months for the conventional structure and 2 months for the roof membrane.

Section 1:250

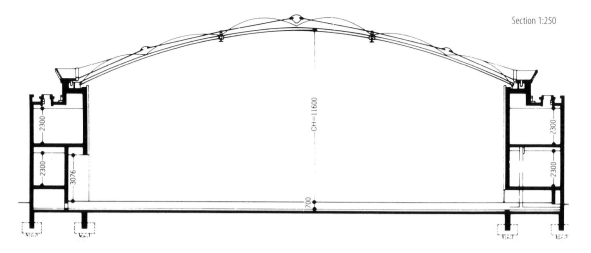

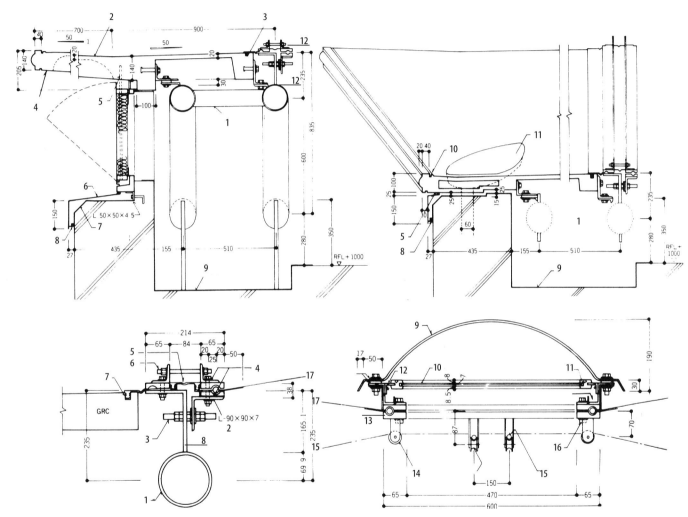

Eaves section: 1 edge arch (steel tube 2 x Ø 89.1 mm), 2 edge element (precast GRC element with waterproof coating), 3 butyl seal, 4 waterproof acrylic urethane coating (2x), 5 drainage tube, 6 aluminium flashing, 7 waterproof coating, 8 urethane seal, 9 urethane coating, 10 groove, 11 drainage, 12 membrane connection (see detail)

Membrane details: 1 arch structure (Ø139.8 x 6.0 mm), 2 steel angles (L 90 x 90 x 7 mm), 3 bolt (stainless steel M20), 4 neoprene padding, 5 neoprene zipper profile, 6 bolt (stainless steel M12), 7 butyl seal, 8 steel plate web, 9 acrylic dome light (Ø 600 mm), 10 wire glass, 11 corrosion protection for wire glass edge, 12 drainage opening, 13 membrane clamping and prestressing ring, 14 roller (stainless steel Ø 36 mm) 15 steel cable (Ø 10 mm) with nylon sheathing, 16 clamping bolt for prestressing ring (M 12), 17 membrane

View of entrance area with roof canopy and bracing longitudinal arch

Festival Theatre for the International Eisteddfod, Llangollen, Wales

Design

Although the architectural brief required a modern building, the design team wanted the theatre to fit into its local environment despite or even because of its membrane roof. From there an architectural concept emerged in which traditional and modern building materials were joined and where old structural forms appear side by side with modern technology. Thus materials like slate, nature stone and wood prevail in the different building parts, while steel and membranes form the connection between them.

Concept

Two natural stone walls rise from the landscape. They form the central walls of the core building, proceed right across the building and beyond it into the landscape. The roof membrane covers the space between the stone walls; at the same time it forms a multi-use hall (lobby) and the arena and thus draws visitors into the respective areas. The ancillary rooms are treated like a two-storey lean-to against the stone walls – with a simple reinforced concrete structure, a timber roof covered with natural slate, painted blockwork walls inside and the staircase towers at either end dressed in stone on the outside. The landscaping around the building is mounded, which further emphasises the effect of the building nestling in its local environment. This is further reinforced by the structure rising out of the ground, by the front entrance to form a porte cochère, and by the buttresses of the arena supporting the main arch. For the architects it was important to keep the structural principles simple and understandable and thus to make the building less daunting.

Use

The building consists of two areas each with a different character. The permanent and the temporary arena (performance areas) serve for events with large turnouts like for example the Eisteddfod. The core building is intended for small sporting events, exhibitions, conferences, theatre performances and receptions, together with the respective ancillary rooms. It was erected in more conventional construction; it has, however, also a membrane roof over its multi-purpose hall, which can accommodate three badminton courts or theatre seating for 350. The core building has full thermal insulation, central heating and forced ventilation. In contrast both arenas are membrane-covered areas without heating and with natural ventilation. Both parts

Client
Clwyd County Council

Architects
General + membrane structure: D. Y. Davies Associates

Structural engineering
Membrane, structure + substructure: Atelier One

Specialist consultants
HVAC: Synergy
Lighting: Equation Lighting Ltd.
Drainage: Generer and Partners
Quantity Surveyors: Venning Hope

Contractors
General contractors: Tilbury Douglas Construction Ltd.
Subcontractors Membrane: Clyde Canvas Goods & Structures Ltd.
Steel: Westbury Tubular Structures plc, Sheetfabs Nottingham Ltd.

Completion
1992

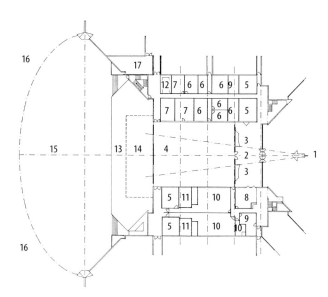

Ground floor plan 1:1000: 1 entrance, 2 lobby, 3 display, 4 multi-purpose hall and stage, 5 toilets, 6 office, administration, 7 assembly room, 8 reception office, 9 store, 10 locker room, 11 shower, 12 kitchen, 13 stage, 14 store room under the stage, 15 auditorium: permanent arena (1990 seats) 16 expansion: temporary arena (4085 seats), 17 plant room

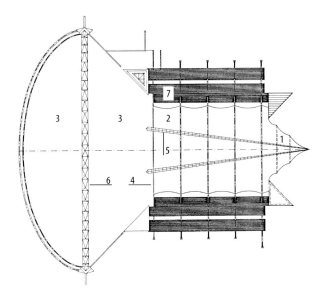

Roof plan 1:1000: 1 roof canopy, 2 membrane roof over multi-purpose hall, 3 membrane roof of the permanent arena, 4 planar steel truss arches of the multi-purpose hall, 5 bracing longitudinal arches (three-chord truss), 6 main arch of the arena (three-chord truss), 7 slate roof over core building with timber roof structure

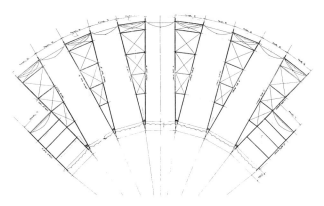

Plan; temporary arena 1:1000: Membrane roof with planar, twin aluminium hip truss frames

Permanent arena with main arches and
core building in the background

Overall view

can be operated and used independently, but are connected with each other.

The arena is a large, covered area with a roof structure which dominates the design. It also lends the building a strong and dramatic identity making it both a landmark and civic statement for Llangollen. It is composed of two parts. The temporary arena is a grassed area, sloping gently towards the stage where a further 4000 seats can be accommodated. It is only used and covered for the Eisteddfod and others large-scale events. The permanent (main) arena with a covered plan area of 2250 m^2 contains a stage with stage lighting; it is essentially a flat area which can accommodate 1800 seats, or can be used for five-a-side football, basketball, volleyball etc. or house trade fairs and exhibitions under its roof. Part of the seating is intended to be housed on demountable tiers, to improve the sightlines.

The buildings had to be constructed within a year, i.e. in the interval between two Eisteddfod festivals.

Structure

Multi-purpose hall | The roof of the multi-purpose hall in the core building is composed of a prestressed tensile roof membrane between curved steel trusses. The arch supports lie on two opposite wings of the core building. The longitudinal bracing, which at the same time provides an additional architectural feature, consists of two three-chord truss arches above the roof of the multi-purpose hall, which rise from a common footing in front of the entrance area. They brace the transverse truss arches laterally and simultaneously carry the cantilevering transverses joists of the roof canopy.

The roof over the heatable multi-purpose hall is executed with a triple-layer membrane. The outer membrane is composed of PVC-coated polyester fabric with a tensile strength of 7.45/6.4 kN/50 mm (warp/weft) and a weight of 1300g/m^2 corresponding to type IV (according to German grading). Below it lies a lightweight, double PVC-polyester membrane with an intermediate glass fibre thermal insulation quilt (d = 100 mm). For reasons of fire protection the inner membrane is made of flame proof, silicone-coated glass fibre fabric.

All three membrane layers are connected by an aluminium boltrope extrusion bolted onto the chord members. A plate flashing above the upper chord and a thermal insulation below form the outer space-enclosing elements and weather protection. A vertical membrane apron forms the connection to the longitudinal walls.

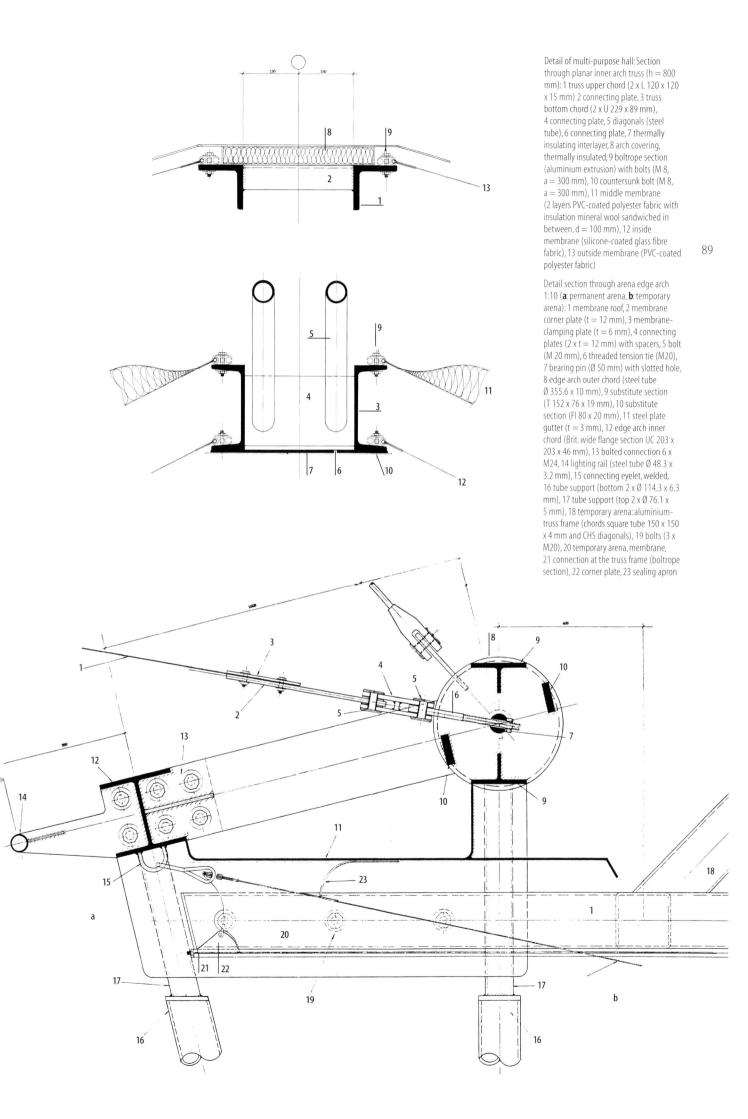

Detail of multi-purpose hall: Section through planar inner arch truss (h = 800 mm): 1 truss upper chord (2 x L 120 x 120 x 15 mm) 2 connecting plate, 3 truss bottom chord (2 x U 229 x 89 mm), 4 connecting plate, 5 diagonals (steel tube), 6 connecting plate, 7 thermally insulating interlayer, 8 arch covering, thermally insulated, 9 boltrope section (aluminium extrusion) with bolts (M 8, a = 300 mm), 10 countersunk bolt (M 8, a = 300 mm), 11 middle membrane (2 layers PVC-coated polyester fabric with insulation mineral wool sandwiched in between, d = 100 mm), 12 inside membrane (silicone-coated glass fibre fabric), 13 outside membrane (PVC-coated polyester fabric)

Detail section through arena edge arch 1:10 (**a**: permanent arena, **b**: temporary arena): 1 membrane roof, 2 membrane corner plate (t = 12 mm), 3 membrane-clamping plate (t = 6 mm), 4 connecting plates (2 x t = 12 mm) with spacers, 5 bolt (M 20 mm), 6 threaded tension tie (M20), 7 bearing pin (Ø 50 mm) with slotted hole, 8 edge arch outer chord (steel tube Ø 355.6 x 10 mm), 9 substitute section (T 152 x 76 x 19 mm), 10 substitute section (Fl 80 x 20 mm), 11 steel plate gutter (t = 3 mm), 12 edge arch inner chord (Brit. wide flange section UC 203 x 203 x 46 mm), 13 bolted connection 6 x M24, 14 lighting rail (steel tube Ø 48.3 x 3.2 mm), 15 connecting eyelet, welded, 16 tube support (bottom 2 x Ø 114.3 x 6.3 mm), 17 tube support (top 2 x Ø 76.1 x 5 mm), 18 temporary arena: aluminium-truss frame (chords square tube 150 x 150 x 4 mm and CHS diagonals), 19 bolts (3 x M20), 20 temporary arena, membrane, 21 connection at the truss frame (boltrope section), 22 corner plate, 23 sealing apron

Interior view: Multi-purpose hall

Tradition meets innovation

Main arena | The roof of the permanent main arena consists of two membrane elements on both sides of the large main arch with its span of 60 m. The first part spans from the last arch truss of the multi-purpose hall to the main arch and is connected intermittently along the wing walls. The second roof membrane extends from the main arch to the edge of the arena, where the membrane is connected to a horizontal arch truss, raised on tubular struts. This edge arch has an effective depth of 1000 mm and consists of an inner chord with a parallel lighting boom and an outer tubular chord. Where the membrane is connected the chord tube is cut and replaced by a statically equivalent, open made-up section with two circular plate diaphragms to receive the membrane pin.

Along the main arch the membrane is supported by a continuous, garland-shaped clamping strip with edge boltrope and with an attached membrane apron, running parallel to the arch and suspended via a threaded bar tie from the bottom chords of the truss arch. For ventilation the clamping strip is widened in parts through a welded inlay piece and protected against the rain by a flashing.

The membrane edge consists of a garland cable with double corner plates, which are fastened with threaded bar ties and pin-joined to the steel edge structure. A welded plate gutter runs around the entire membrane under the garland edge to drain off the rainwater.

Temporary arena | This roof will only be erected for the summer months. The structure consists of pairs of planar, aluminium hip frames, connected with each other through a bracing of tubular struts and tension diagonals. Six of these frame pairs and two planar single-edge frames in front of the wing walls at the arch supports carry the temporary membrane roof. It is connected at regular intervals along the bottom chord tubes of the truss frames and anchored via an edge cable to the bottom edge of the truss frame and by earth nails to the ground.

Roof canopy | The single layer roof canopy is composed of three membrane elements. They are connected to steel angles which are carried as cantilevers by the bracing longitudinal arches. At the ends of the cantilevering angles the membrane is fixed at a membrane tip plate, to which the edge cable (Ø 19 mm) is also anchored via a threaded fitting. The tip plate is pretensioned towards the outside by a threaded round steel bar running in two guides parallel to the cantilever. An eyelet and notch facilitate the installation. At the top the roof canopy is tied down by two cables to the arch base.

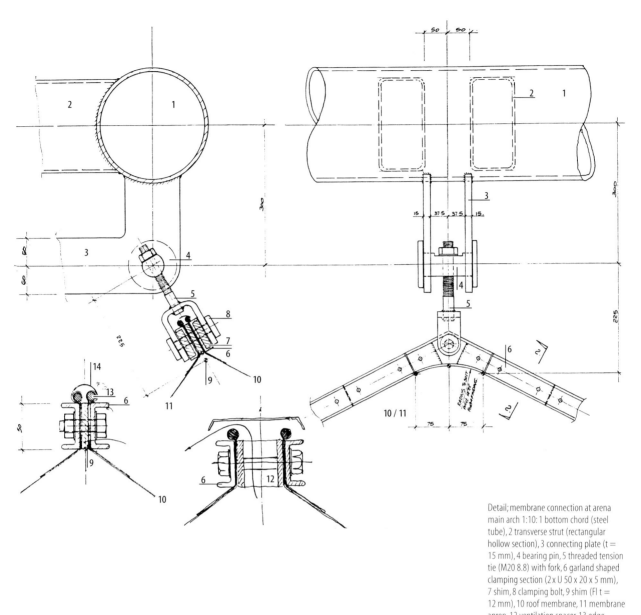

Detail; membrane connection at arena main arch 1:10: 1 bottom chord (steel tube), 2 transverse strut (rectangular hollow section), 3 connecting plate (t = 15 mm), 4 bearing pin, 5 threaded tension tie (M20 8.8) with fork, 6 garland shaped clamping section (2 x U 50 x 20 x 5 mm), 7 shim, 8 clamping bolt, 9 shim (Fl t = 12 mm), 10 roof membrane, 11 membrane apron, 12 ventilation spacer, 13 edge boltrope, 14 membrane covering strip

View from the harbour

Mobile Theatre for Buddy Holly Musical, Hamburg, Germany

Client
Buddy KG

Architects
General: Urban Project GmbH, Klaus Latuske
Membrane structure: IPL Ingenieurplanung Leichtbau GmbH

Structural engineering
Membrane, structure: IPL Ingenieurplanung Leichtbau GmbH
Buildings: Sellhorn Ingenieurgesellschaft

Specialist consultants
HVAC: Sellhorn Ingenieurgesellschaft
Acoustics: Sellhorn Ingenieurgesellschaft
Project management: Sarnafil AG (primary structure), Sellhorn Ingenieurgesellschaft and Urban Project GmbH (secondary structures and site supervision)

Contractors
Membrane: Sarnafil AG
Steel: Philipp Holzmann AG
Airconditioning: S + H Nolting

Completion
1994

Design

The 'Neue Metropol Theater' in Hamburg commissioned the construction of a mobile theatre which was to be used for the first time for a musical on the life-story of Buddy Holly, the Texan rock star who died in 1959 in a plane crash. The mobile theatre, with a textile roof structure covering a floor area of 500 m^2, was constructed in 1994 within the space of six months. The roof construction itself was designed and planned within only eight weeks. The theatre is located in the free port area opposite the St. Pauli pier.

The membrane roof had to cover an auditorium with 1400 seats, with a main and two side stages as well as a two-storey lobby in the front part of the long, rectangular building.

The lobby was closed with a cable-stabilised glass facade, curved in plan, which offers a splendid view over the city and harbour.

The 1400 velvet-upholstered chairs are arranged under a dark blue ceiling; by skilfully arranging the 11 lower and 15 upper rows and the six boxes, no seat is further than 25 m away from the stage. Each row of seating is raised 12 cm above the one in front and thus offers a good view of the three stages.

The double-layer skin enveloping the building consists of an outer and an inner membrane, which serve simultaneously as weather protection, soundproofing and thermal insulation. The outer membrane is translucent, while the inner one is opaque, so that theatre performances are possible during the day. Although the structure is described as a mobile arched hall, it is conceived as a permanent building. After two years of service it is scheduled to be disassembled at the end of 1996 and set up again in a different location.

Heating, ventilation, air-conditioning | The spaces between the inner membrane and the upper edge of the facade as well as the internal membrane space along the roof edge are closed by a double layer sealing apron. A 40 to 80-cm thick airspace between inner and outer membrane provides thermal and acoustic separation. A specially developed ventilation system allows effective and energy-saving air conditioning. Warm inner air is drawn into the membrane space, improved by adding fresh air and returned again into the auditorium. This method of circulating the air allows a multiple use of the already warmed inside air and thus leads to considerable energy savings. Separately controllable exhaust fans in the apex of the hall complete the ventilation system.

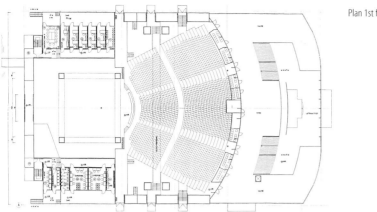

Plan 1st floor 1:000

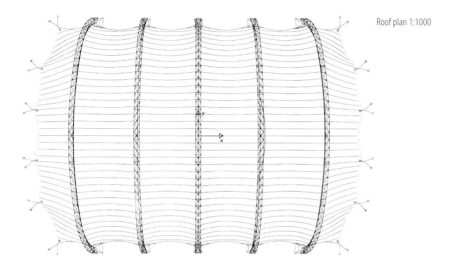

Roof plan 1:1000

In the summer the double membrane works as a cold roof. For natural ventilation of the membrane space both air inlets are available from the interior as well as from outside. The space is ventilated via roof vents in the upper part of the membrane.

Structure

The primary structure consists of five welded pin-supported steel truss arches (at 15 m c/c), in the shape of a three-centre arch, with a span of 55 m and an effective depth of 1.5 m, which are stabilised in longitudinal direction through prestressed steel cables. The middle arches are vertical, the outer ones are slightly tipped to the outside which gives the building its distinct appearance. Each arch consists of five sections for ease of transport.

At each end of the building six tripods are placed, where both the stabilising cables and the edge cables of the roof membrane are attached. The building envelope consists of a double layer roof membrane of PVC-coated polyester fabric. The outside membrane is supported along the arches with a garland cable and intermittently connected to the arch bottom chord. The inner membrane is suspended in garlands under the arch. At the edge along the facade, both membranes are anchored via edge cables in arches and tripods and are prestressed from there.

Loads | For snow load 0,75 kN/m² were assumed, according to the German DIN standard and to its location Hamburg. The design wind load is q = 0,8 kN/m² (v = 129 km/h) according to DIN standard.

Arches and tripods | The two-pin arches are braced in longitudinal direction by four prestressed cables, which join the arches; they are anchored at the building ends at the above-mentioned tripods. The self-supporting roof membrane spans between the arches. The three-chord tubular

Cross section 1:500

truss of mild steel St37 consists of five short, transportable sections, joined by high-strength friction grip bolts in a circular pipe flange: top chord tube Ø 244.5 x 8 mm, bottom chord Ø 323.9 x 16 mm, diagonals Ø 101.6 x 4.5 and Ø 114.3 x 7.1 mm, and the transverse struts in the middle arch Ø 48.3 x 5 mm.

A tripod consists of a compression member inclined at 15° to the vertical and of two diverging tension members with a steel bar tie fixed above to carry the forces from the stabilising cables.

Arches and tripods are protected by a zinc dust undercoat and by a double top coat; fittings and fixings are galvanised.

Membrane roof | The roof consists of four saddle-shaped membrane parts spanning between the five arches and of two free-form end membranes as transition from the arch form to the polygonal edge formed by edge cables.

For the existing brief this form has some distinct advantages:

A rectangular plan can be covered without difficulty.

Through the arch construction a large building height and associations with circus tents are avoided and the loads are still transferred into the ground in an efficient way.

Due to the subdivision of the roof membrane into panels and their pointwise connection at the arches the membrane installation is also relatively simple.

To connect them with the arches the outer membrane panels are equipped with a continuous steel edge cable (Ø 16 mm). Rounded corner plates are connected with forked turnbuckles via gusset plates welded to the arch bottom chord tube. With a continuous membrane apron clamped to the chord tube the membrane is closed and sealed to the outside.

The inner membranes of two adjacent bays are connected underneath the arch along the parting line by two angle sections and edge boltropes. They are connected with the bottom chord tube by an almost vertical membrane strip lying in the arch plane, with edge cable (Ø 12 mm), corner plates and forked turnbuckles.

At the tripods the outer and inner membranes are connected to a complex, welded, multilayer mast head structure:

The corner plates of the inner membrane are connected via an adjustable eye bar which runs in a welded-on tube sleeve with connection plates welded onto the edge mast (compres-

Longitudinal section 1:500

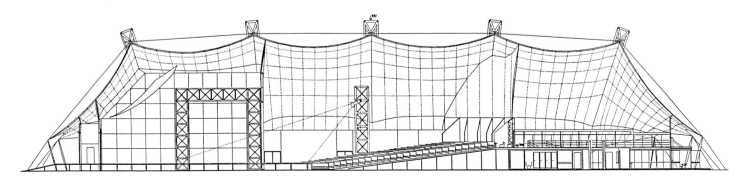

Interior

sion strut). The corner plates of the outer membrane are joined via a bolted connection to two plates, welded onto the tripod head.

The stabilisation cables (spiral strand Ø 28 mm) are fixed by means of a bent steel plate welded onto the tripod head. This carries the tension forces via two round bar sections, lying above the tripod guys, into the system point of the guy structure (tension strut). This way additional bending moments in the tension guy are largely avoided.

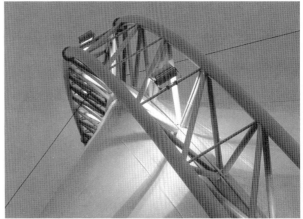

Membrane connection at arch and tripod

Detail view: Truss arch with membrane connection and stabilising cables

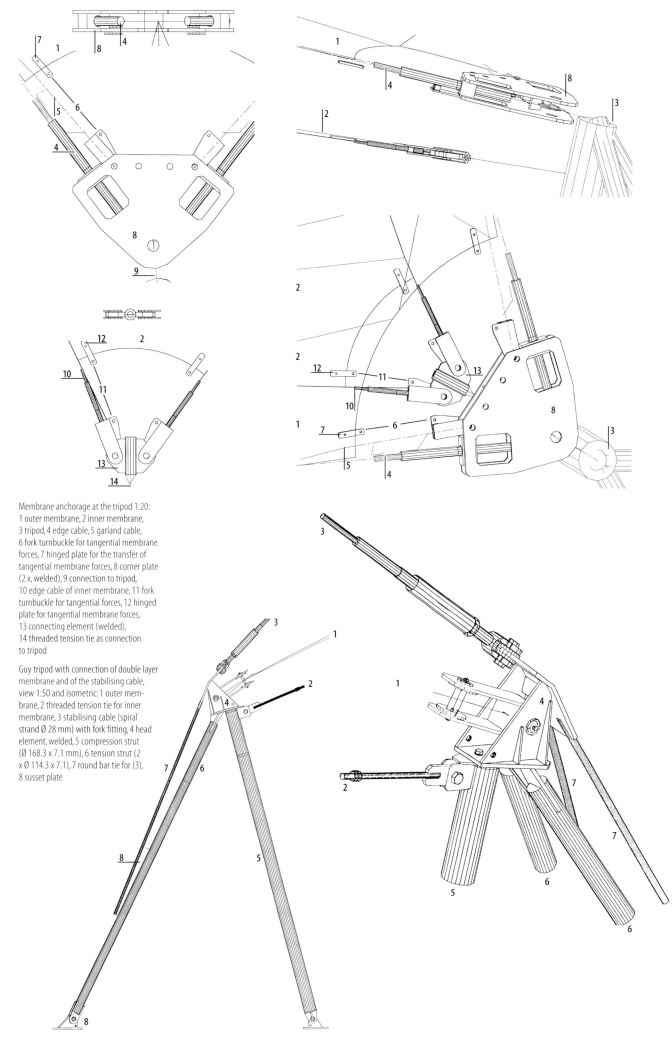

Membrane anchorage at the tripod 1:20:
1 outer membrane, 2 inner membrane,
3 tripod, 4 edge cable, 5 garland cable,
6 fork turnbuckle for tangential membrane forces, 7 hinged plate for the transfer of tangential membrane forces, 8 corner plate (2 x, welded), 9 connection to tripod,
10 edge cable of inner membrane, 11 fork turnbuckle for tangential forces, 12 hinged plate for tangential membrane forces,
13 connecting element (welded),
14 threaded tension tie as connection to tripod

Guy tripod with connection of double layer membrane and of the stabilising cable, view 1:50 and isometric: 1 outer membrane, 2 threaded tension tie for inner membrane, 3 stabilising cable (spiral strand Ø 28 mm) with fork fitting, 4 head element, welded, 5 compression strut (Ø 168.3 x 7.1 mm), 6 tension strut (2 x Ø 114.3 x 7.1), 7 round bar tie for (3),
8 susset plate

Aerial view with entrance tent and Galileo Tower

National Research Exhibition 'Heureka', Zurich, Switzerland

Design

Concept, brief | The exclamation of the Greek philosopher, scholar and engineer Archimedes, 'Eureka' (I found it), is the name and motto of a research display, which was shown in 1991 in the 'largest tent of the world' in connection with the celebrations to mark the 700th anniversary of the Swiss Confederation.

In a round 'tent' and in different annexes (in the wooden Galileo tower, the Heureka-polyhedron and a central 'micro cosmos-macro cosmos' pavilion), a general display was presented about the current state of science and research and its future development.

For the jubilee celebrations of the Swiss Confederation an appropriate cover was to be provided with approx. 17 000 m^2 plan area and for a period of just over 5 months. The membrane structure should fit harmoniously into the building site at the Allmend Brunau in Zurich and it should be quick and economical to erect and to dismantle. Its temporary nature should be taken into account in the structural design in terms of its required strength and service life.

Situation | The Allmend Brunau lies in the Sihltal on the outskirts of Zurich. On one side it borders onto a road and motorway, in the west rises the Uetliberg, while the ground on the remaining sides is flat. The subsoil is sufficiently firm: Under an upper layer of topsoil lies river gravel with a thickness of approx. 12 m, with moraine gravel underneath.

Final scheme

The free-form membrane roof, created by the Swiss sculptor Johannes Peter Straub as a stretch fabric model with eight high points, was further processed by the membrane engineers through modern CAD methods. The central exhibition area is housed in a ring tent, which consists of 8 separate but connected textile structures. The plan shape of a single tent is a circle of 25 m or 31.45 m radius; the two larger tents being situated in the middle, on both sides of the main entrances formed in the membrane by entrance arches on the central axis.

The height of the masts is between 32 m and 42 m; the interior terrace structures have two and three stories, which results in a total exhibition area of about 30 000 m^2. Four viewing platforms at the mast heads are accessed by spiral staircases around the masts and through elevators and are connected with each other through suspension bridges.

Client
Züricher Forum, Zurich

Architects
Design: Johannes Peter Straub, sculptor, Zurich
Exhibition architect & project coordination: P. Angst-Obi, Zurich

Structural engineering
General and steel structure: Plüss + Meyer, Lucerne
Membrane, structure and substructure: IPL-Ingenieurplanung Leichtbau, Radolfzell

Contractor
Project management: General contractor and membrane: Sarnatent, Sarna
Kunststoff AG, Sarnen
Subcontractors steel: Baltensberger, Höri (ZH)

Completion
1991

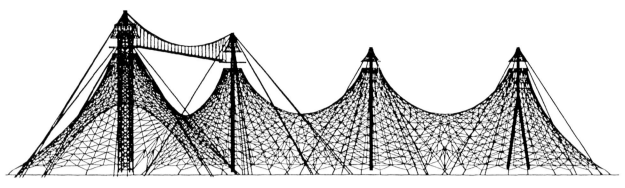

Side view; computer perspective

Street view

Structure

The primary structure consists of 8 vertical steel tube masts, some of which are connected by suspension bridges. In addition to their normal function in the membrane roof, the masts serve as

- support for the viewing platforms at approx. 30 m in height,
- newel posts for the spiral stairs,
- elevator shafts and as
- support for the suspension bridges.

Thus their horizontal displacement under load had to be limited to a maximum of 15 cm. Depending on the specific use of interior space and mast type the masts have different structural systems: In tent 1 the mast was executed as an A-frame trestle (steel tube Ø 812.8 x 16 mm), to create a column-free space for an auditorium. The masts carrying the spiral staircases were detailed as tube masts (Ø 1016 mm) fixed at their base, while masts 4 and 5 are equipped with elevators and constructed as lattice trusses (2.9 x 3.3 m, Ø 323.9 x 12.5 mm) encastréed at the base.

Load assumptions | Despite its temporary nature an evenly distributed snow load of 0.40 kN/m^2 and an uneven load of 0.5/0.3 kN/m^2 were assumed in the analysis.

The terrace structures, bridges and staircases were designed for a live load of 4 kN/m^2.

Wind loads were determined in accordance with an expert analysis of wind loading; the design wind speed was determined as 147 km/h.

Suspension bridges | The twin cables of galvanized steel wire are carried over saddle supports at the masts and anchored in the ground in tension foundations. At masts 4 and 5 a sliding support was introduced so that only vertical forces are carried into the masts. The link bridges are made from two transverse rectangular hollow sections, connected by the bolted-on sheet metal walking surface. To carry the horizontal wind forces they act as horizontal plate girders.

To be able to reuse the suspension bridge constructions, they were executed with standard loads, corrosion protection and surface finish for a regular design life.

Foundations

The foundations for compression loads were executed as reinforced concrete pad foundations, the tension foundations are micropiles (injection anchors), where a steel tube sleeve with a Gewi bar tie is drilled into the ground. Subsequently the tube sleeve is partially pulled and the anchor pressure grouted.

Membrane | The roof membrane consists of PVC-coated polyester fabric Sarnafil S, type 4 with a tensile strength of 7.5/6.5 kN/5 cm; the seams have a width of 100 mm and are high-frequency welded. For cost reasons and because of its short design life no additional protective surface lacquer was used, which resulted in pollution traces becoming visible even during its lifetime. The structural high point membrane sheds its loads into 8 masted high points and is anchored at the edge through edge and guy cables.

The galvanised spiral strands (edge cables Ø 20 mm to 32 mm, guy cables Ø 26 mm to 32 mm) are equipped with swaged fork or eyelet fittings. In general the membrane corners are tied down directly; only at the sides of the entrance arches edge masts were used. At the membrane corners the edge cables are connected with the guy cable via single pins and a triangular twin corner plate with tube spacers.

The edge cables run in membrane sleeves. Tangential forces in the membrane are carried into the corner plate through a 2.5-m long clamping plate chain and a turnbuckle.

The edge masts have triple guys and consist of a central steel rod (Ø 85 mm) with conical reinforcement plates (t = 20 mm) welded on. They are equipped with a semi-circular end resting in a spherical cut-out in the steel base plate. The guys of the edge masts are anchored in the tension foundation by means of a cylindrical socket, a two-piece transom and a U-shaped, threaded prestressing bar anchored to Gewi sleeves and bars.

At the high point the membrane is connected by clamping plates to a stiff steel tube ring. The ring is suspended from the mast ring by 8 short steel cables. At the guyed masts these hanger cables are arranged vertically, so that essentially only vertical forces are entered into the masts. At the other masts the hanger cables run diagonally in the direction of the mast tip, where they are covered by a membrane mast hat.

The six smaller tents (1-3, 6-8) have a simple, funnel-shaped mast hat, which is clamped at the top to a tube ring and tensioned at the bottom by a garland cable with corner plates using belt straps pulled over a tube ring. The two larger tents with the viewing platform at the mast head have a two-part hat with similar structural details.

Project management

The commission for the scheme design went directly to the membrane contractor. As some parts of the building were sponsored, no general contractor was appointed: the manufacture and supply of the steel structure, cables and footbridges as well as the foundations were commissioned directly. The project management was carried out by Sarnafil AG as membrane contractor.

Production and installation | The cutting patterns were produced electronically at the office of IPL and transmitted electronically to the membrane manufacturers where they were plotted directly onto the membrane, which was subsequently cut manually. Likewise the patterns for the steel plates were plotted in full scale and optically read at the steel contractor to control the plate cutting.

The primary structure was erected and guyed, with masts up to 42 m high, including stairways and terraces; the footbridges were assembled, then the entire membrane surface of 23 000 m^2 was spread out at the ground and connected from 27 individual membrane panels through clamping plate joints and lifted within half a day using a specially developed lift arrangement. Subsequently the mast head covers were assembled, and mast 1 was converted to the aforementioned A-frame trestle. The dismantling proceeded in reverse sequence. Due to the possibility of a second application the membranes were dried again at the works.

Roof membrane with edge cable arch and direct guy

Edge mast: 1 membrane, 2 edge cable with eyelet fitting in edge cable sleeve, 3 clamping plate-chain for transfer of the tangential forces from the membrane edge, 4 turnbuckle, 5 edge cable at entrance opening with fork fitting in edge cable sleeve, 6 edge mast, (round Ø 85 mm with welded-on plate stiffeners t = 25 mm), 7 guy cable with fork fitting and threaded U-bolt, 8 base plate with welded-on ring sleeve made from short steel tube, 9 steel ball joint, 10 Gewi anchoring

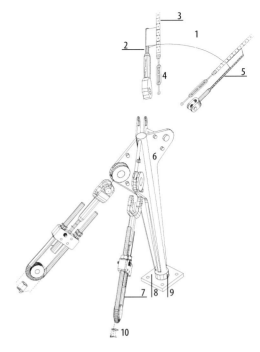

Direct guy: 1 membrane, 2 edge cable with eyelet fitting in edge cable sleeve, 3 clamping plate chain for transfer of tangential forces from the membrane edge, 4 turnbuckle, 5 corner plates (2x), 6 guy cable with eyelet fitting, 7 threaded U-bolt, 8 two-piece transom, 9 cylindrical socket, 10 roller, 11 Gewi anchoring, 12 connecting element

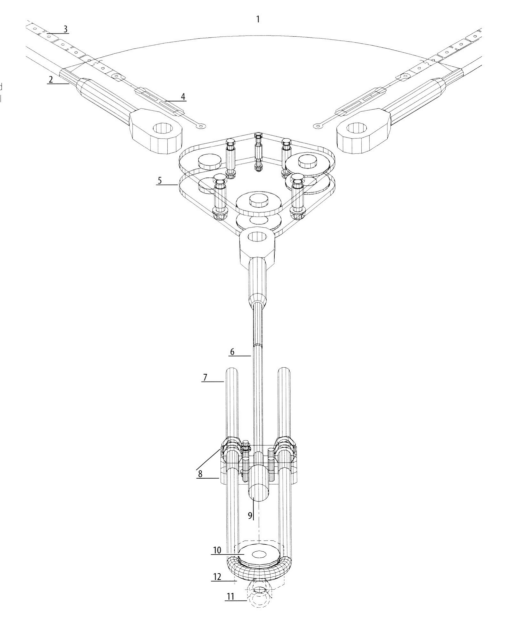

 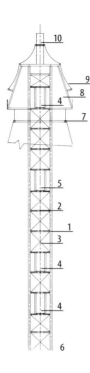

High point detail at viewing mast with suspension bridges and mast hat

Main mast (H = 42 m) with viewing platform and elevator 1:500: 1 steel tube (Ø 323.9 x 12.5 mm), 2 steel tube (Ø 152.4 x 10 mm), 3 round bar diagonals (Ø 50 mm), 4 lift door, 5 emergency door, 6 top of foundation, 7 membrane high point ring (Ø 7.0 m) with vertical suspension, 8 viewing platform, 9 membrane hat with webbing straps, 10 upper mast part (steel tube)

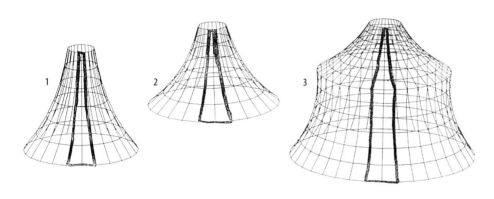

Mast head cover: 1 + 2 funnel-shaped mast hat, 3 two-part hat at viewing platform

Detail at the lower edge: 4 membrane hat, 5 edge cable, 6 clamping plate (2x), 7 bar eye, welded on, 8 clamping bolt, 9 webbing strap, with tensioning device, 10 tube for the anchoring of webbing straps

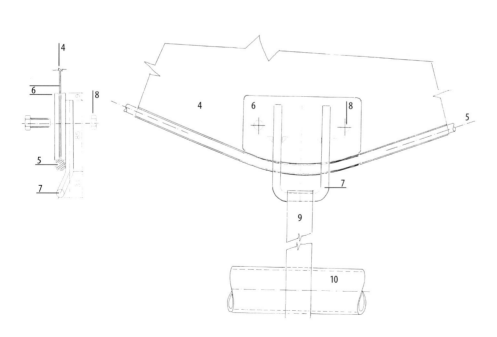

Diagonal view

Pneumatic Hall 'Airtecture', Esslingen-Berkheim, Germany

Design

With this test structure of an exhibition hall the company Festo, an internationally active company in pneumatic components, presents presumably the very first air-supported building with a cubic interior and with a supporting structure consisting primarily of air-inflated elements.

The appearance of the new structure deliberately creates a new image and design, different in construction and appearance from existing air halls. Its design also reflects the advanced technological background and purpose of the hall; in its colour scheme grey is predominant, with accents of the company's colour blue and metallic aluminium, which creates a cool and friendly image in line with the company's range of colours.

Situation, dimensions | Inside the hall has a floor area of 375 m^2 and a height of 6 m, resulting in a total volume of 2250 m^3. The covered exterior dimensions are 800 m^2, with an external height of 7.2 m. With a total dead weight of 7.5 kg/m^2 the structure is considered to be light. The hall was erected on the company ground; other applications are intended.

The two end walls of the hall consist each of two L-shaped elements, leaving a free space between them measuring 3 m x 6 m. These two wall openings are used as entrances and above one of them the ducts for supply and exhaust air enter the hall. A 4 m high, transparent PVC membrane serving as a roll-up gate is connected to a steel frame independent of the air structure.

Structure

The hall consists of approximately 330 individual air-inflated elements, most of which come under the six categories of wall components, transparent window cushions, Y-shaped columns, roof beams with translucent, intermediate membranes and pneumatic tension elements (so-called 'muscles'). They all have different volumes and internal pressures.

The load-bearing structure of the exhibition hall includes 40 Y-shaped columns and 36 wall components with parallel faces along both longitudinal sides. 72 000 distance threads per square metre hold the double-layer walls in place at an operating pressure of 0.5 bar. The slits between the opaque wall components are filled with transparent cushions made of Hostaflon ET as window elements; slide locks allow easily installation.

The load-bearing structure has many new features. These include among others the use of double-wall-fabric as a load-bearing element, flame-inhibiting elastomer coatings, a new trans-

---

Client
Festo KG, Esslingen

Architects, Design
Festo KG, Esslingen, Corporate Design/ Pneumatic Structures; Rosemarie Wagner, Axel Thallemer, Udo Rutsche
Working drawings of substructure: Architekt Jaschek, Stuttgart

Structural engineering
Festo KG, Esslingen, Corporate Design / pneumatic structures
Foundations: Ingenieurbüro Willi Rathgeb, Esslingen

Specialist consultants
HVAC: Festo KG, Esslingen

Contractors
General contractor and pressure management: Festo KG, Esslingen
Fabric structures: DSB Deutsche Schlauchboot, Eschershausen
ET-foil cushions and erection: Koch Konstruktive Membranen GmbH & CO KG / KOIT High-tex, Rimsting

Completion
1996

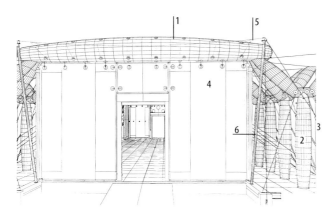
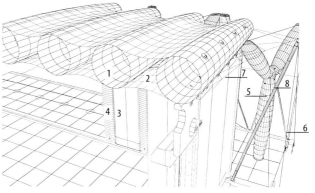

lucent Levaprene coating (EVA = ethylene-vinyl acetate coating), and last but not least, computer-controlled pneumatic tension elements ('muscles') and the concept of a building actively responding to environmental forces. The building was designed for a snow load of 50 kg/m^2 and a simultaneous wind speed of 80 km/h.

Roof structure | The horizontal air-inflated girders, so-called "airbeams", carry the roof loads. Their maximum diameter in the middle of the span is 1.25 m tapering to 0.75 m at the supports. The airbeams are supported at the junction of the wall components and the Y-shaped columns along the sides of the hall and are connected with them by textile webbing. The airbeams are pretensioned by their internal pressure thus forming a stiff structure.

Translucent membranes connect the upper and lower sides of the airbeams. These intermediate membranes are stabilised by a low pressure vacuum. The alternation of positive (500 mbar) and negative (2.5 mbar) pressure produce a stiff roof diaphragm.

Bracing | The columns, which in plan are arranged in saw-tooth-pattern, together with one pair of wall elements, form a triangulated structure carrying vertical loads, for example snow, with the main proportion being carried by the walls. Horizontal wind loads are carried by the space frame consisting of pneumatic tension elements called 'muscles', pneumatic columns and wall elements.

The end girders on top of the end walls are braced laterally by a stainless steel boundary cable which carries the wind load in this area and transfers it via galvanised steel struts and two guy cables into the foundations. The edge cable also serves to resist horizontal forces from the low-pressure intermediate membranes. All cables are connected to vulcanised patches with D-rings and cable terminals.

'Muscles' | With increasing wind speed, the 'muscles', supplied by increased inflation pressure from the pressure management, react with increased tension. The computer-controlled pneu-

View; Y-columns, 'muscles', wall and transparent window cushion elements

End wall with edge girder and edge cable with tubular steel mast and guy

Perspective view in longitudinal direction; 1 roof beam, 2 pneumatic Y-column, 3 pneumatic 'muscles', 4 wall element, double-wall fabric with spacer threads, 5 edge cable, 6 tubular steel strut with guy

Perspective section through roof structure; 1 roof beam, 2 low pressure intermediate membrane, 3 wall element, double wall fabric with spacer threads, 4 ET-foil cushion window 5 pneumatic Y-column, 6 pneumatic 'muscles'; 7 edge cable, 8 tubular steel strut with guy

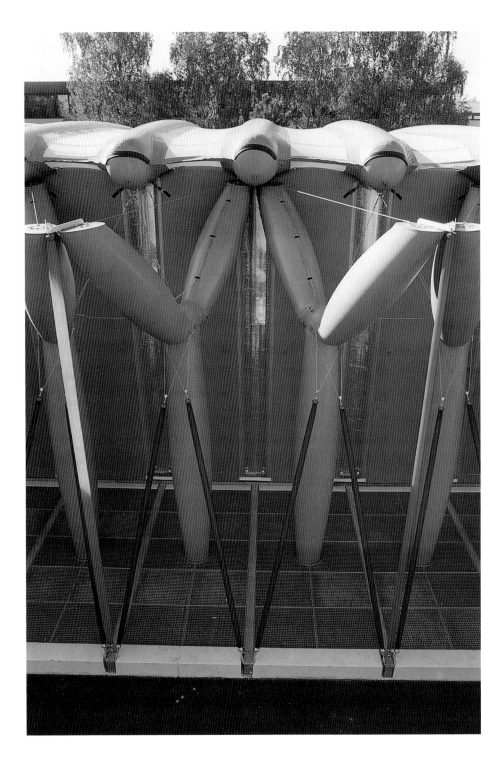

Side view; Y-columns, 'muscles', wall and transparent window cushion elements

matic muscles consist of a special manmade fabric with internal silicone hose; tensions can be controlled by varying their inflation pressure. The tensile forces of the outer fabric are in equilibrium with the air pressure on the internal hose membrane. The air pressure in the muscles can be varied between 0.3 and 1 bar thus regulating the axial force.

Foundations | Narrow steel rails are connected to concrete strip foundations; their half-pipe shells support the wall elements. This provides a counterbalance against wind uplift forces. Grating is used as a floor inside the hall, supported on steel joists fixed to the foundations at 1 m centres and 50 cm above the gravel bed. To emphasise the lightness of the structure the Y-columns rest on steel I-beams. Grating is also used as flooring in the exterior.

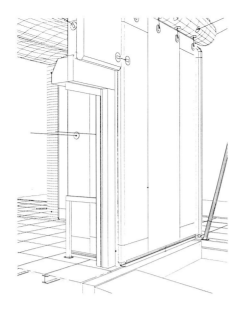 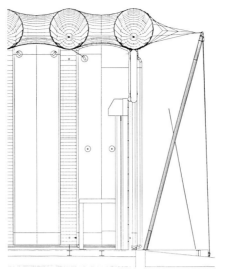

Perspective view and section at gate

Pressure management, controls

A speciality of this lightweight structure is its unusual structural system and its ability to actively respond to wind and snow loads. Relevant environmental data are supplied by a weather station on the building.

To control the inflation pressures, all air-inflated elements are grouped into 10 almost identical sections, each of which is controlled by a valve terminal through proportional valves. The actual pressure in the elements is monitored by sensors. If an excess pressure builds up in any one element, this can be vented by means of pressure relief valves. The pneumatic control elements for each section are housed on mounting plates, installed in the floor space between grating and gravel. All plates are connected to a ring pipe, encircling the hall foundations and feeding the valves. The air pipes are routed from there to the pneumatic elements easily accessible through wall slits.

As a high-level control system a programmable control unit is used to communicate with the individual terminal slaves (i.e. the local control units mentioned above, which are dependent on the main control unit) and with the weather station via a data cable (field bus). The system calculates the necessary pressure values in the individual air elements according to the actual climatic conditions.

Heating, ventilation, air conditioning

In addition to the natural ventilation through the two doors, an air conditioning system is provided. The conditioned air is distributed via 2 textile supply air-supply ducts, suspended from the ceiling. Gravel fill under the grating is used as heat storage; light-coloured Rhine gravel keeps infrared absorption to a minimum. Radiators in the air space between grating and gravel bed heat the hall in winter. In summer a cooling system is available to cool the hall. The double layer walls of the exhibition hall promise improved thermal insulation compared with conventional membranes.

Transport, assembly

The air-inflated chambers made from textile membranes can be folded into a container of about 14 m in length and may be hauled quickly and efficiently to another site due to its low total weight of 6 t.

Entrance area with arch and concrete edge and A-shaped edge support trestles

Suntory Pavilion, Expo '85, Tsukuba, Japan

Design

The exhibition in this pavilion stood under the mottos 'Bird life' and 'Our future: tomorrow'. The exhibition was intended to present new science and technology and to give an impression of the shape of our future society. A building in keeping with these themes was to be designed, with fluent lines and curved surfaces, resembling a gently rising hill or a bubble arising from the Earth.

Plan, use

The inside area of the pavilion was divided into three zones: an exhibition area, an auditorium with projection room and numerous ancillary rooms. The world's largest projection screen so far (26 m x 36 m) was placed in the centre of the pavilion. Around it on the first floor the various ancillary rooms were arranged. On the second floor the exhibition areas are located in the same ring-shaped plan area.

Structural concept

For the architect the most important design task of the first design phase was to cover the gigantic projection screen with an organically formed roof. After long consideration and investigation of different schemes a roof structure was chosen which follows a reverse catenary. In the beginning a wooden gridshell was explored, however, this would have required the solution of a number of tricky problems, like for example a decision on the required safety factors for the curved wood grid structure or satisfying the demanding fire protection requirements.

Instead it was decided to use a similar form with a structure made from planar rigid frame elements. From arch truss elements a "naturally" curved roof shape was formed corresponding to the architectural concept. Under the circumstances this appeared to be the best solution for the construction of the desired shape and for carrying the considerable wind loads.

Fire protection | For fire protection smoke vents and short, vertical flameproof drop walls in the roof area had to be provided as separations for the fire compartments of the automatically opening smoke vents. In the projection room a fire shutter was required to close this area against the auditorium in case of a fire. Also for fire protection in the lower membrane area up to 4 m above the floor a roof membrane of PVC-coated glass fibre fabric had to be used.

Client
Suntory Ltd.

Architects
General and membrane structure:
Keizo Sataka + Institute of International Environment

Structural engineering
Membrane, structure and substructure: Masao Saito and T.I.S & partner

Contractor
General contractor and substructure: Takenaka Corp.
Membrane and steel: Taiyo Kogyo Corp.

Completion
1983 –1985

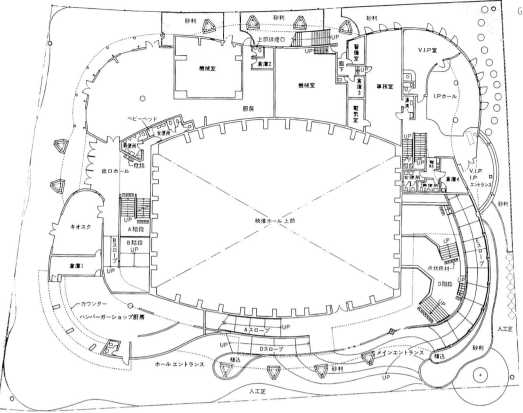

Ground floor plan 1:500

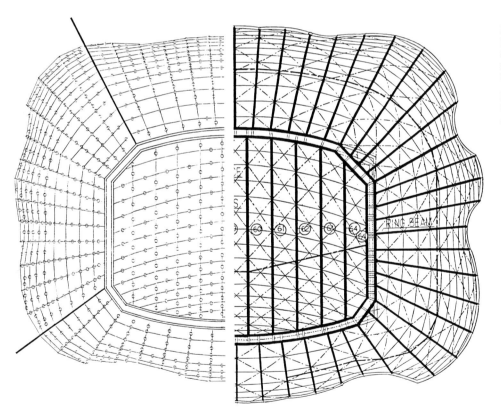

Left: Roof membrane 1:500 with high points and erection joints. In the ring-shaped outer area the membrane is subdivided by four erection joints (fat lines).

Right: Roof structure 1:500: arch and bracing through tension ties and a compression ring

Edge structure

Interior view of the entrance area: roof structure and edge beam

Hanging chain model from jewellery chain for the form-finding of the catenary shape

Section at the outer roof area 1:100: 1 arch beam at the auditorium roof (Ø 267.4 x 7.1 mm), 2 tie bar (Ø 25 mm), 3 roof membrane with high points, 4 smoke vents, 5 arch beams at the exhibition roof (Ø 267.4 x 9.3 mm), 6 tie bar (Ø 25 mm), 7 roof membrane with high points, 8 compression strut (Ø 101.6 x 4.2 mm), 9 arch edge, 10 lacing against edge tube Ø 25 mm, 11 plate gutter, 12 ring beam (steel tube), 13 wall cladding (gypsum plasterboard), 14 projection screen, 15 Vierendeel wall structure, 16 edge beam, 17 sheet metal connection plate for membrane and decorative cover strip, 18 lacing against steel edge tube Ø 25 mm, 19 cover strip (400 x 454 mm), 20 gutter over entrance area, 21 reinforced concrete substructure

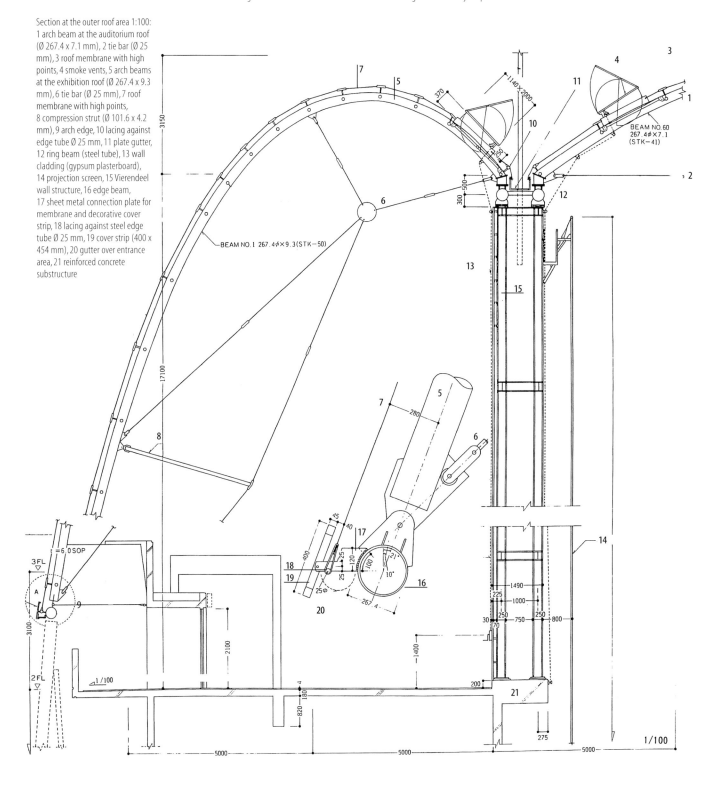

Structure

The structure has a conventional reinforced concrete substructure, exposed concrete walls, floor slabs as well as strip and pad foundations, reaching from the foundations up to the exhibition area in the second floor. The structure above consists of three parts:
- the auditorium walls,
- the roof of the auditorium, and
- the ring-shaped roof of the exhibition areas.

Wall | The walls of the auditorium are set up as cavity walls; they consist of two rolled steel sections connected by gusset plates, which form a Vierendeel strut. They are clad on both sides with gypsum plasterboard. A twin ring beam made from steel tubes lies in the upper wall edge.

Roof | The auditorium roof consists of 10 truss arches in the shape of a catenary, which in plan are arranged parallel to each other and which are pin-supported on a ring beam. The roof of the exhibition area has the same structural system, however, with a radial arrangement of the arch trusses.

The planar arch trusses consist of steel tubes (Ø 267 x 9.3 mm in the outer roof and Ø 267 x 7.1 mm above the auditorium) with round bar tension ties (Ø 25 mm) and compression struts (CHS Ø 101 x 4.2 mm). They are pin-supported on the edge and ring girders.

The tension ties are equipped with welded end plates (t = 19 mm), which are connected via short connecting plates (t = 19 mm) and with two bolts each to the chord tube and to plates welded onto the strut head (t = 19 mm). At the tensile nodes the round bar diagonals are connected via short connection plates of similar construction (also with two bolts each) to circular steel plates (t = 19 mm).

According to the type of loading, part of the tie suspension will go slack and therefore not carry load, which was taken into account in the structural analysis as a change of the structural system.

In the area of the compression struts the arches are connected at right angles to the arch plane by steel tube compression members and braced against lateral buckling by tie cross-bracing in the plane of the arch chord.

The straight steel tube joists (Ø 89 x 4.2 mm) are placed in the centre plane of the arch chords and are connected with it via steel plates and machine bolts.

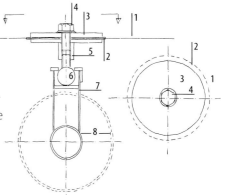

Detail; adjustable high point 1:10:
1 membrane, 2 membrane reinforcement patch, 3 circular clamping plate, 4 connecting bolt, 5 threaded sleeve, 6 ball joint with threaded bar welded on, 7 tube prop, 8 arch beams (max. and min. tube diameter)

Edge structure | In the edge area the roof is supported by three-dimensionally curved edge beams, which shed their load into A-shaped steel trestles standing outside the building.

The A-shaped supports consist of an external steel tube compression strut with a welded steel plate tripod construction. The outer compression member is furnished on top with a conical reduction and a spherical head. There connection strips are welded on to receive the edge beams.

The edge beams consist of two steel tubes connected with truss diagonals. They are bolted to the A-trestles, on the outside with steel plate strips and on the inside using a pipe flange. They carry the arched roof beams, which are connected via a pin connection and welded steel strips to the plate stiffeners.

Along the reinforced concrete edge the arches are pin-connected on steel supports. These supports are fixed onto the horizontal concrete surfaces by means of anchor bolts and a mortar bed.

Membrane | The roof membrane is supported on approx. 1600 high points on the arches and joists. The position of these support points was determined by computer during the form-finding stage, while simultaneously controlling the surface shape to achieve the desired roof form.

The material is PVC-coated polyester fabric treated against surface spread of flame, with a tensile strength of 400 kg/5 cm (warp) and 365 kg/5 cm (weft/fill).

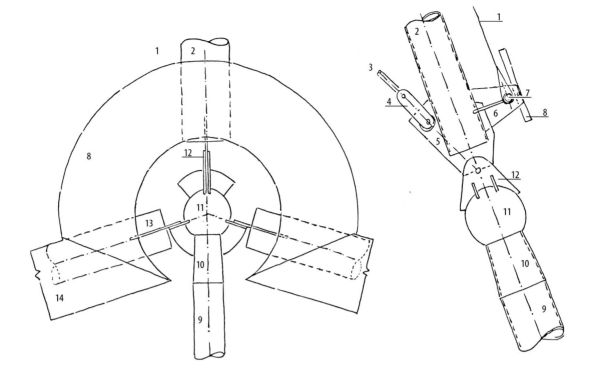

Top view and side view of edge detail 1:20:
1 roof membrane, 2 arch beams, 3 tie bar Ø 25 mm, 4 connecting element, 5 gusset plate, 6 sheet metal connection plate for membrane and cover strip, 7 lacing against steel edge tube Ø 25 mm, 8 cover strip (400 x 454 mm), 9 steel tube compression strut of the A-shaped trestle, 10 conical reduction, 11 steel ball with 12 welded connecting plates, 13 arch edge beam, 14 decorative cover plate in front of edge beam

In the lower membrane area, up to 4 m above the floor, it was replaced for reasons of fire protection by a PVC-coated glass fibre fabric and at the interface welded to the PVC-coated polyester membrane. The areas around the smoke vents are also made of this PVC-coated glass fibre (for the same reasons).

At the high points the membrane is supported and fixed in position by two circular steel plates (Ø 220 mm) with a central machine bolt and reinforced by a membrane patch on the outside and inside. The lower circular plate is welded onto a steel tube spreader, which is welded to arches and joists in a direction perpendicular to the roof surface. Some of the high point props are adjustable in angle and length, so that the membrane can be prestressed and tolerances can be equalised. For this a threaded bar with ball head is located in a welded bearing held at the prop end and screwed into a threaded sleeve. At the other end of the sleeve a machine bolt is screwed in, clamping both circular plates against the membrane.

Along the tubular beams the roof membrane is connected by lacing to a tube (Ø 25 mm) running parallel to the tube beam and connected at the edge beams by short welded plate strips. In front of the lacing a decorative cover strip is fixed, which hides it from view. This strip is bent and cut into a ring shape around the ball heads on top of the A-trestles, emphasising the support. At the entrances a gutter is provided under the lacing, shedding the rainwater to the side.

At the concrete edges the membrane is also fixed via a covered lacing to an uninterrupted tube section (Ø 25 mm), which is fastened to the reinforced concrete by short plate strips. In the upper roof area the membrane edges are detailed in a similar way; here a membrane apron covers the lacing connection to the Ø 25 mm tube and at the same time sheds the rainwater into a box gutter, lying between the two ring beams on top of the wall.

In the ring-shaped outer roof area the membrane is subdivided by four site joints. The four roof panels are again connected through lacing to a steel tube (Ø 25 mm). The joint is covered and sealed from the outside by a membrane strip.

Detail view with membrane corner, high point cover of transparent polycarbonate sheets, ET-foil roof with aluminium arch and cable bracing

Trade Fair Stand for the 'Automechanika', Frankfort, Germany

Together with the trade fair architects of the office WNS Design in Wiesbaden the project 'A Filling Station of Tomorrow' (30 x 20 m in plan) was to be developed on behalf of a carwash plant manufacturer. A futuristic design and details using new materials, new ideas – innovative, visual and functional – was requested for the archetypal concept of a 'filling station'.

The stand was to be demountable and should contain a novel filling station with space for several cars. The first station of this type was erected in the open-air at the IAA-Fair in Frankfort, with the possibility of further uses by the client for information and promotions.

Design

The architects and the membrane technology experts from IF worked together on novel building materials, particularly concentrating on fabric-reinforced plastics and foils and on developing the membrane detailing. The detail development owes much to ship building, and in this case especially to the design of the large sailing yachts on Lake Constance. The fittings and fixings, cables and masts of these yachts are part of a technology which has developed over centuries and become standard in daily construction tasks. In this field examples and models for the solution of many detailing problems like cable anchorings and prestressing and adjustment solutions could be found, and these ideas were then modified for the kind of specific, one-off production which is still a necessity in the construction industry.

Structure

The roof membrane in the main structure is self-supporting. In the interior of the main roof there are three steel tube masts with high point rings and with a conical, clear transparent cover; these masts carry the roof.

Along one side the roof has a straight edge. There, short steel tube sections (Ø 26.9 mm) are placed into a membrane sleeve, cut out intermittently with semi-circular cut-outs, and anchored at keys welded on along the bending stiff edge made from a steel I-section lying on its side.

The service cabins inserted under the edge cable arches of the main roof membrane consist of tubular aluminium three-centre arches fixed at their base with a transparent foil stretched over it.

Client
Kleindienst Waschanlagen, Augsburg

Architects
General: WNS Design GmbH, Wiesbaden
Project architect: Wolfgang Nikolaus Schmidt
Membrane structure: Horst Dürr, IF Ingenieurgemeinschaft Flächentragwerke

Structural engineering
Membrane, structure and substructure: IF Ingenieurgemeinschaft Flächentragwerke, Dipl.Ing. R. Dinort

Contractors
Membrane and steel: Stromeyer & Wagner GmbH, Constance
Foil roof: Colux, Singen

Completion
1994

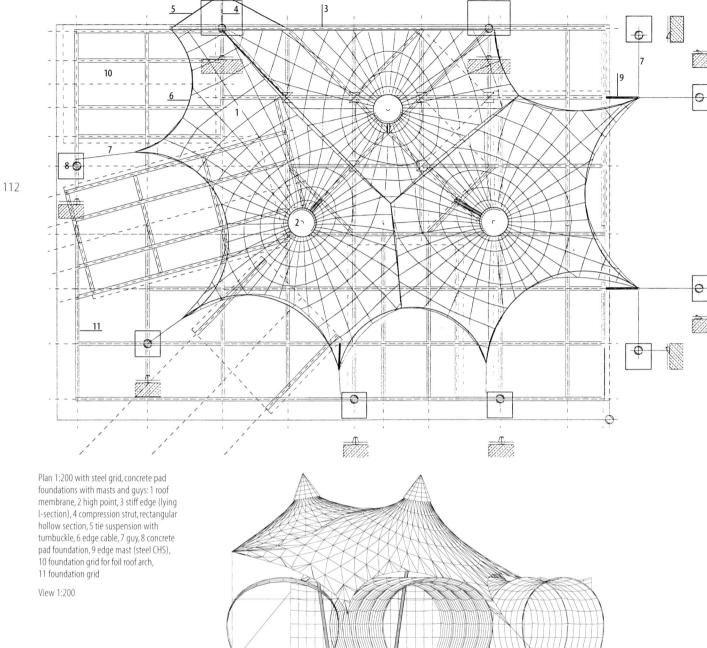

Plan 1:200 with steel grid, concrete pad foundations with masts and guys: 1 roof membrane, 2 high point, 3 stiff edge (lying I-section), 4 compression strut, rectangular hollow section, 5 tie suspension with turnbuckle, 6 edge cable, 7 guy, 8 concrete pad foundation, 9 edge mast (steel CHS), 10 foundation grid for foil roof arch, 11 foundation grid

View 1:200

View during assembly

Foundations

Due to its construction on the trade fair, special foundations like tension foundations or gravity anchors with large concrete volumes in the ground were not possible. Therefore a steel grid was chosen to spread the loads, the weight and stiffness of which is sufficient, together with the weight of the simple pad foundations and of some external tension ties, to prevent uplift of the structure under wind.

Membrane | The main membrane consists of PVC-coated polyester fabric, type I, and the foil cover of the arches of clear transparent ET-foil (Hostaflon).

Along the edge the membrane is reinforced and held by an edge cable. It runs in membrane pockets. An additional edge webbing transfers the tangential membrane forces and allows an independent adjustment to be able to equalise production tolerances. The edge cables are equipped at the end with steel thimbles and swaged aluminium sleeves and connected with a standard shackle to the corner plate. The edge webbing is sewn up at the end with the triangle ring and joined to the corner plate via an adjustable connection.

To this end a threaded rod is bent into a triangular 'ring' and welded up; the threaded bar runs through a tube sleeve welded into the triangular plate, where it is anchored with a tensioning nut. The corner plate is connected with the edge mast and its guy structure via an adjustable eye bar, consisting of an eye plate with a slotted hole to receive the associated shackle and of a threaded bar which is welded on. Here, too, the threaded bar runs in a tube sleeve welded into the corner plate and anchored with an tensioning nut. At a direct guy without edge-mast the eye bar is replaced by a standard shackle passing through a hole in the corner plate, to which the guy cable is attached via a steel thimble.

Foil roof | The foil cover has an edge boltrope welded in, which is inserted into a curved aluminium boltrope extrusion, similar to that used in large frame tents. The arch structure is braced in a longitudinal direction by stainless steel cable crosses with thimbles and fork turnbuckles and with steel tube struts, fastened to anchor plates bolted to the arches.

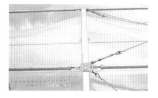

Detail view: Cantilever at I-section edge with horizontal guy

Bracing detail with arch, compression member and tension diagonals with turnbuckle

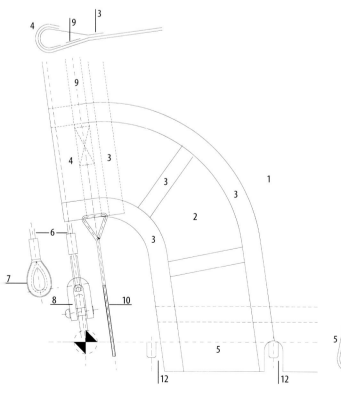

Membrane corner at the stiff edge 1:10: 1 membrane, 2 corner reinforcement, 3 HF-weld, 4 edge cable sleeve, 5 membrane pocket for steel tube anchorage, 6 edge cable with 7 thimble and 8 standard shackle, 9 webbing sewn up, 10 webbing tensioner, threaded bar, 11 tube pieces for anchoring at 12 keys, welded into the stiff edge section (not shown)

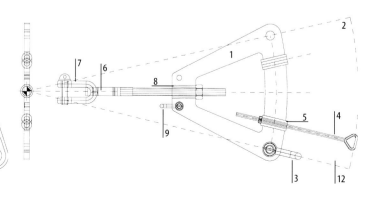

Membrane corner 1:10: 1 corner plate, 2 edge cable, 3 standard shackle, 4 webbing anchorage, 5 tube sleeve, 6 adjustable eye bar with 7 standard shackle, 8 tube sleeve, 9 shackle as assembly aid

Side view with cylindrical centrepiece and collar connection, air condition plant containers and foil window

MEDIADROM Mobile Air Hall for 'ZDF' TV, Germany

Design

The origins of this project go back some way. The second German television channel, ZDF, and Mediadrom GmbH were first interested in the concept of a provisional, but then still stationary pavilion for erection at the Munich trade fair site. During the scheme design phase the pavilion was conceived as a pneumatic structure, i.e. as air-supported. As the ZDF ultimately wanted to also set up the hall in the centres of other large cities, for example next to the Brandenburg Gate in Berlin and beside Cologne Cathedral, and to use it for spectacular music events, three essential requirements were added:
 – The hall had to be mobile, i.e. easy to erect and to dismantle.
 – It had to be transparent from the outside, so that people could look in.
 – It should have good acoustic qualities.
 – Lighting and acoustical equipment of approx. 30 t had to be housed in the hall.

Concept

In co-operation between the architects and the membrane engineers the following solution emerged.

To enable the air hall to be moved quickly from one place to the next, the stationary solid concrete foundations were replaced by mobile, water-filled concrete troughs. All parts had to be designed to be transportable and storable, and very precise schedules were required to be able to erect and dismantle the hall within a week. The entire hall has a transport volume of 60 pick-up trucks.

The requirement to show from outside what's happening inside, was partly fulfilled by the translucency of the chosen material (PVC/PES). During performances at night the spotlights project reflections and shadows onto the hall membrane. As a further measure the sides of the larger hall were opened in the bottom part by the installation of three large transparent foil windows (6 m x 30 m) which permit visual communication between inside and outside. Telecasts therefore are not only watched by the guests present in the Mediadrome but also by many spectators who gather in front of the large foil windows.

Naturally the ZDF gave high priority to acoustics; by means of suspended acoustic sails and acrylic glass reflectors the required high-grade acoustics with concert quality could be achieved. Since it was not possible to attach 30 t of lighting and sound equipment to the pneumatic hall skin due to the resulting high point loads, it was decided, especially with regard to the accom-

Client
Mediadrom GmbH together with ZDF

Design/architect
General: WNS Wolfgang Nikolaus Schmidt Design GmbH, Wiesbaden
Membrane structure: Structural engineering, membrane, structure and substructure: IF Ingenieurgemeinschaft Flächentragwerke, Constance

Specialist consultants
HVAC: Nolting KG
Acoustics: Dr Schraepel, Dr Blum

Contractor
Membrane and steel: Polynederlands BV, Holland

Completion
1995

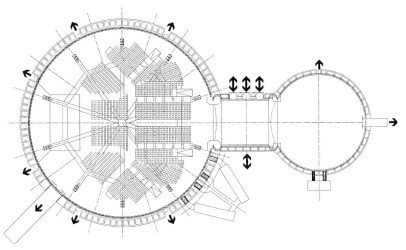

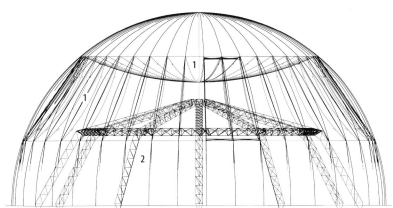

Plan 1:1000 with large hall (with emergency exits, air conditioning units and truck lock), centrepiece (with main entrance) and small hall (with forklift lock and blower)

Section through large hall 1:500 with acoustics sails (1) and technology table (2)

modation of air conditioning and ventilation ducts, to install a separate technology table in steel construction.

Plan, use | The plan form of the large hall (with a floor space of 1950 m^2) is circular, at Ø 51 m, which is joined through a short, cylindrical centrepiece (floor space 190 m^2) with the small hall (floor space 500 m^2) at Ø 26 m. Over it the roof arches in the form of two hemispheres joined by a cylindrical membrane.

The main entrances are situated in two containers at the cylindrical centrepiece, with three revolving doors on one side and one revolving and six escape doors on the other side. The large hall has to have seven emergency exits, according to the building regulations, whereby it has a maximum capacity for approx. 3000 people. In addition a forklift and a truck airlock were installed, the latter with a length of 19 m.

The covered plan area is 2700 m^2; 1800 seats can be installed at the Mediadrome; without seating it has a maximum capacity of 3000 people.

Structure

The structure is an air hall according to DIN 4134 and was designed with respect to the guidelines for movable buildings. The membrane domes are prestressed through the inside air pressure and transfer their loads into the gravity foundations. A blower provides an air pressure of 30-50 mm water gauge; which is controlled by the wind speed and increased automatically if required.

At the transition to the cylindrical centrepiece a membrane collar is installed on each side, which accommodates the displacements and movements of the two different membrane structures meeting. To allow a movement of the hall under load relative to the rigid structures of truck and forklift air lock they were also connected with a collar. The frames of the emergency exits, however, are connected to the main membrane without collar and therefore had to be designed to resist the applied membrane forces.

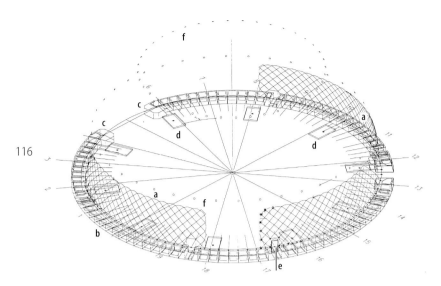

Isometric of large hall: a foil window with cable net and garland edge, b precast concrete foundation element, c special foundations for collar/ edge cable, d technology table foundations, e emergency exit, f connection points for acoustics sails (main membrane not shown)

Interior view with acoustics sails and technology table

Foundations

The mobile foundations (1.35 m x 2.20 m, h = 1.20 m) are trough-shaped precast concrete elements, which are filled with water. By this measure the transport weight could be reduced by 1/3 from originally 900 t, as a compromise between the conflicting requirements of a low transport weight on the one hand and high ballast weight and small volume in the installed condition on the other. The forces are carried almost vertically into the foundations so that the available weight is optimally utilised.

Roof membrane | For the large hall a PVC-coated polyester fabric type IV was used while for the small hall a type II proved sufficient. The membrane windows are made from ET foil (with 200 μm thickness); to stiffen the high membrane force of the large hall, it was reinforced by a cable net with a mesh width of 1.0 m x 1.0 m.

Since various unusual building materials were used, i.e.
– PVC-coated PES fabrics type II and type IV, as well as
– ET foil for the windows,
a proof of the suitability of these materials was required according to the German building code through a process of consent for the individual case, i.e. the suitability of the materials had to be demonstrated with regard to their structural and fire-resistant material characteristics.

The roof membrane is connected to the p.c. foundations via short tube pieces inserted in a welded membrane pocket; the tube clamps made from steel flat are bolted onto a Halfen rail cast into the concrete. A lateral displacement of the tubes is prevented through a welded-on tube flange. The space is sealed through an inner membrane apron, welded onto the main membrane and lying loose in the lower part the building. Any air losses occurring are replenished by the blowers.

Transparent membrane window | For clear transparent membranes only a limited number of foils are available and these generally only have a limited tensile strength. In addition a material classified as 'difficult to ignite' was required. The material chosen was a combination of a difficult-to-ignite ET foil reinforced by an outer steel cable net with a mesh width of 1 x 1 m. The main disadvantage of the foil material is that it would be destroyed after approx. 2 to 3 erections through creasing, unless additional protective measures were taken. To avoid this wear and tear, the hall is equipped for installation with an opaque window of PVC-coated polyester fabric, which after inflation is replaced under inside pressure by the cable net and the ET foil.

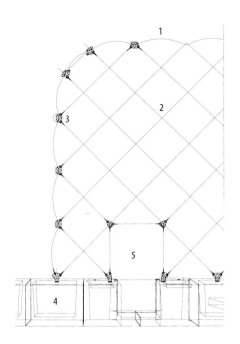

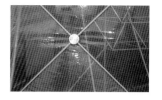

Detail of foil windows 1:100: Garland edge (1) of the main membrane, cable net (2) with connection plates (3) and precast concrete foundation elements (4) with emergency exit (5)

Foil window with steel cable net, emergency exit and precast concrete foundation elements with drain opening

Foil window with steel cable net nodes

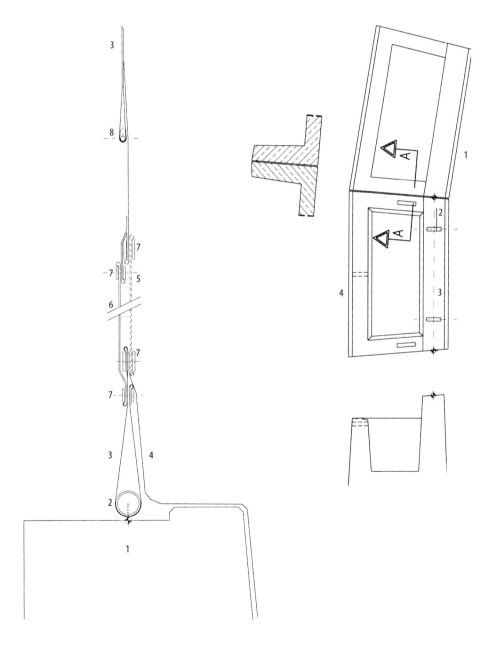

Section through foil windows 1:5:
1 precast concrete foundation element, 2 steel tube in membrane sleeve, 3 main membrane, 4 sealing apron, 5 internal, opaque window membrane of PVC-coated polyester fabric, 6 external ET-foil, 7 clamping strip connection of the window membranes for connecting it with the main membrane (3), 8 garland edge with 'endless' edge cable in membrane sleeve

Precast concrete foundation element (1) with water filling; 2 steel flat clamp with Halfen rail, 3 connection tube for membrane connection (axis), 4 overflow pipe (1:50)

The foil structure between the meshes of the steel cable net transfers the inside air pressure through tension forces in the membrane. The steel cable sheds its load via corner plates into the garland edge of the main membrane and is thus part of the dome membrane. The garland edge is composed of an 'endless' edge cable, running in a membrane pocket. At the connection points of the net, cables corner plates are installed outside and inside of the membrane joined together by bolts; the net cables are pin connected via a separate connection plate.

The foil and the opaque window membrane from PVC-coated polyester fabric are connected with the main membrane by a staggered clamping strip connection with boltrope.

Installations

Acoustics: Wall and ceiling sails and acoustical reflectors | To achieve the required, high-grade interior acoustics with the unfavourable hemispherical form, 40 triangular point-connected roof and wall sails were installed and suspended from the roof membrane.

The acoustics sails are composed of a PVC-PES-net fabric, covered with a 6-cm thick mineral wool blanket. Because of the net the sound is not reflected at the inner surface, but penetrates the acoustics package, becomes reflected at the second, outer surface of the insulation package, runs a second time through the insulation layer and is so optimally (i.e. doubly) dampened. The wall sails work the same way.

Through the installation of hard acrylic glass panels at the technology table the sound arriving in the lower areas is unevenly reflected and thus modifies the reverberation time.

Air-conditioning | The space is ventilated with 90 000 m^3/h and is fully air conditioned, the required air-conditioning plant is housed in two containers set up beside the halls.

Floor | The floor is an adjustable floor system with Mobilplex floor slabs and a permissible live load of 10 kN/m^2.

Assembly

The erection and dismantling takes only a week due to a specially trained, 50-strong installation crew. Delivery and placing of the foundations takes place according to a precise schedule. The foundation elements are assembled directly from the truck, other components are unloaded and distributed with forklifts. The spreading of the membrane can only be done by hand, i.e. through employment of as many workers as possible. As a protective measure slip foils are used, for example when pulling the membrane over the foundations.

When inflating the hall while it rains there is always the danger of pooling, which must be pumped out if necessary; under strong wind conditions the hall cannot be inflated.

The foil window is installed as follows: The hall is inflated with the help of the opaque window of PVC-coated polyester fabric; then the inside air pressure is lowered for the installation of ET-foil and cable net. The ET-foil is placed in position and clamped on; then the stainless steel cable net is assembled and fixed from the outside. The opaque window membrane is disconnected at the bottom and at the side, rolled up to the top and remains hanging there as a curtain (and is available as such if required). The inside pressure is raised again to the standard level. Installation of the foil window extends the total setting-up time, but as the ET foil was the only material available which was classified as 'difficult to ignite' this disadvantage was inevitable.

Side view

Exhibition Building 'Big Wave', Hiroshima, Japan

Design

The concept of this unorthodox and spectacular exhibition building for the Sea & Island Expo '89 in Hiroshima sprang from the wish to graphically express Hiroshima's particular geographical and climatic situation by the sea: as an easy, playful expression of a message for world peace in this city destroyed by the atomic bomb. Nowhere else is there such a large building, which has the three-dimensional shape of waves. For this purpose the 'Ukiyoe' pictures by the Japanese painter and graphic artist Hokusai were studied and transferred into a built form: the whole range of wave movements and wave forms, restless intermediate areas, the breaking of the waves at the shore and the froth they leave behind.

Two large exhibition areas were to be created: one for a 20-m high, full-scale section model of the flagship of Christopher Columbus, the 'Santa Maria', and one for a large show object smaller in height and including an audio-visual display.

For the representation of the wave form for both areas a pneumatic, air-supported structure was chosen. Hall 1, with a required room height of max. 30 m, was braced by a substructure of curved, cable-braced steel trusses, to avoid the large displacements under wind load common for a steep pneumatic structure. A flat connecting area between the two "waves" is designed as an open, mechanically prestressed tent structure in the form of a high point membrane with inner masts.

Plan

The exhibition hall lies on a peninsula amid an artificial stretch of water with a beach. Nearby passes a funicular cable car. In the long, narrow Hall 1 with elliptical plan the display object Santa Maria and some associated exhibits are located, with seating tiers for the visitors, various storerooms and entrance and exit areas with the necessary airlocks for people and goods as well as access ramps. As a connection between Halls 1 and 2 a lounge area with catering was inserted as a buffer zone, which also contains the plant room for blowers and standby generators. The plan form of Hall 2 is irregular, it resembles a circle with small bulges and results from the outer wave form and from the compact form of the audio-visual display shown there.

Structure

In the area of Exhibition Halls 1 and 2 the structural system is an air hall, i.e. a roof membrane of coated fabric supported and stabilised by inside air pressure.

Client
Sea & Island Expo Associates

Architects
General and membrane structure:
Keizo Sataka + Institute of
International Environment

Structural engineering
Membrane, structure and substructure: Institute of International Environment + Gengo Matsui and Taiyo Kogyo Corp.

Contractor
General: Nomura Co., Ltd.
Membrane: Taiyo Kogyo Corp.

Completion
1988–1989

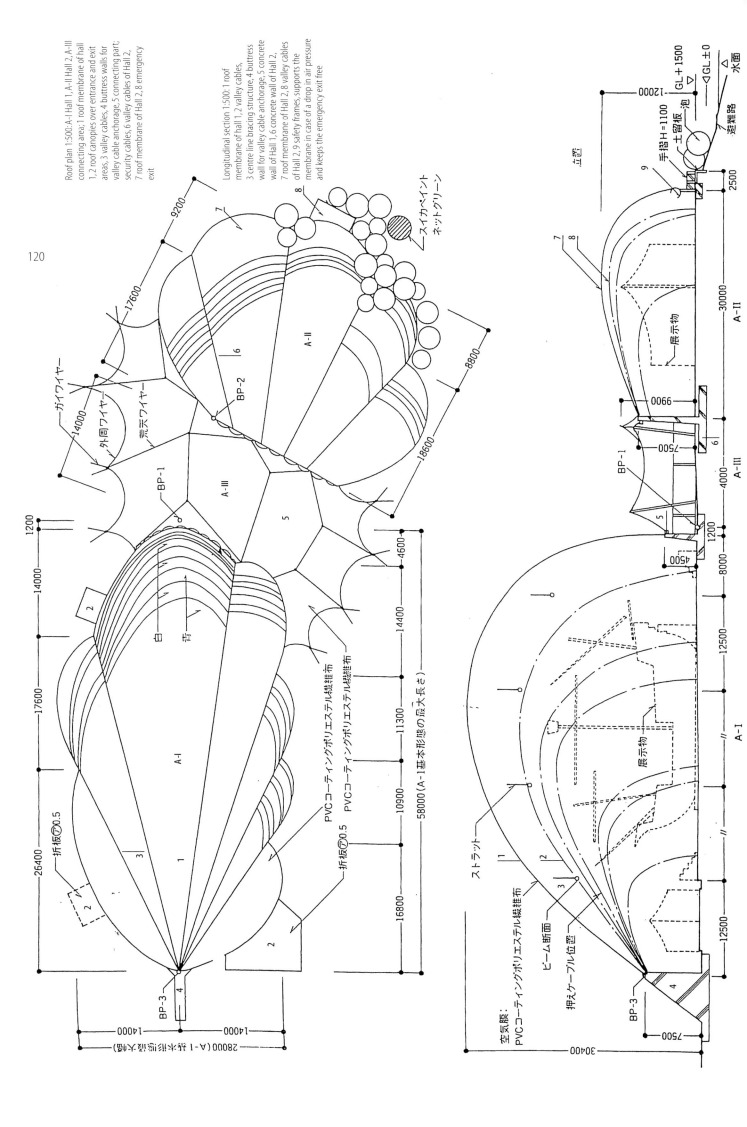

In the connecting area the structure is a mechanically prestressed high-point membrane, which is reinforced on the outside by edge cables and held in position through edge masts and guys. All along the reinforced concrete walls facing the air-supported parts the membrane is fastened to the walls and drained through a gutter there.

Membrane | The roof membrane of the air halls is composed of opaque PVC-coated polyester fabric, likewise the mechanically tensioned, anticlastic (saddle-shaped) membrane of the connecting part. The structural seams are carried out as combination seams, i.e. welded and sewn seams combined and was partly sealed with a external weather-strip, welded over with hot air.

Air halls | The shapes of the air-supported waves can be interpreted approximately as torus or translation surfaces, with the valley cables as directrix.

The air-supported structure is stabilised through inside pressure, which at standard conditions amounts to 20 mm water gauge and can be increased if required up to 40 mm water gauge, to provide the necessary stiffness for extreme wind speeds. (The design wind speed because of possible typhoons is 60 m/sec = 216 km/h).

The individual membrane parts are strengthened through valley cables. Their anchorages in Hall 1 are located in a wall shaped, 7.5-m high reinforced concrete support on the entrance side near the airlocks, on the other side the valley cables are connected at a 4.5-m high reinforced concrete wall. In Hall 2 the valley cables proceed from a 7.5 m high reinforced concrete wall on the one side to anchorages at low level in the strip foundation on the other side. In Hall 1 the garland-shaped valley cables are connected to a substructure of prestressed, cable-braced compression arches. These form a planar, bending-stiff truss as a lateral bracing. The individual trusses have a steel tube bottom chord with bolted flange connections and reinforcement ribs, a top chord from prestressed steel cables, steel tube web members and cable diagonals with turnbuckles. In transverse direction the truss elements are braced laterally with prestressed cable ties, which proceed from the compression chord of one truss to the tension chord of the other.

In the valleys the membrane is joined with a structural valley apron membrane, lying above the valley cable, through sewn and welded membrane reinforcement strips. This valley apron is reinforced at the edge with a thin edge cable (boltrope) and is connected intermittently with steel plate strips, machine bolts and metal eyelets to the valley cable.

At the edge of the air halls the roof membrane is finished with garland-shaped edge cables, running in membrane pockets. The continuous edge cable is anchored point by point through shackles and connecting plates with U-bars cast into the reinforced concrete foundation. The roof membrane continues as a membrane apron across the edge cable and lies loosely on the concrete foundation, where it is pressed on by the inside air pressure. To transfer the tangential forces from the edge cable pockets these are fastened to the corner plates with thin plate strips. The corner plates transfer their loads by means of rounded anchor shackles and cast-in round steel U-bars in the foundations.

Connecting area | The transition between the two wave-shaped air halls is formed by a high point membrane with six inner high points, edge cables and edge masts with guys on two sides. At the remaining sides the roof membrane is fixed onto the reinforced concrete structure of the 'waves'.

The high points are carried by steel tube masts, where the membrane is connected via high point rings. The masts proceed through the high point rings to a point above where they are joined with each other through a system of security cables. These serve to secure the masts or other structural elements against collapse in case the membrane was destroyed through an extreme loading condition. The high point rings are closed by conical covers.

View from a gondola of the funicular cable car

Interior view with model of the 'Santa Maria'

Assembly: Primary structure with concrete wall, buttress wall for valley cable anchorage and scaffolding

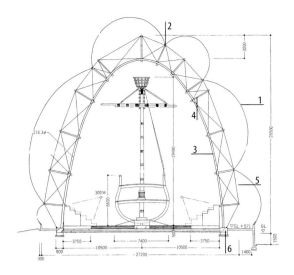

Cross section through large hall 1:500 with ship display and seating platforms: 1 roof membrane, 2 valley cable, anchored in bracing structure, 3 bracing structure: cable-braced arch structure with compression arch (CHS Ø 216.3 mm), 4 strut (CHS Ø 190.7 mm), 5 cable bracing, 6 reinforced concrete strip foundation/tension anchor

Membrane connection at the valley cable and edge detail 1:3, 3: 1 valley cable (Ø 42 mm), 2 steel plate clamp (t = 1.6 mm, b = 50 mm), 3 bolt M12, 4 boltrope (Ø 9 mm), 5 membrane apron, 4-ply, 6 connection eyelet, 7 roof membrane, PVC-coated polyester fabric, 8 cover-/reinforcement strips with combination seam (sewn and HF-welded), 9 edge cable sleeve, 10 edge cable swaged with thimble, 11 connecting bolt for edge cable, 12 corner plate, 13 strip connection for tangential forces from membrane, 14 anchor shackles with bolt, 15 concrete anchorage, cast in round steel U-bar (Ø 22 mm)

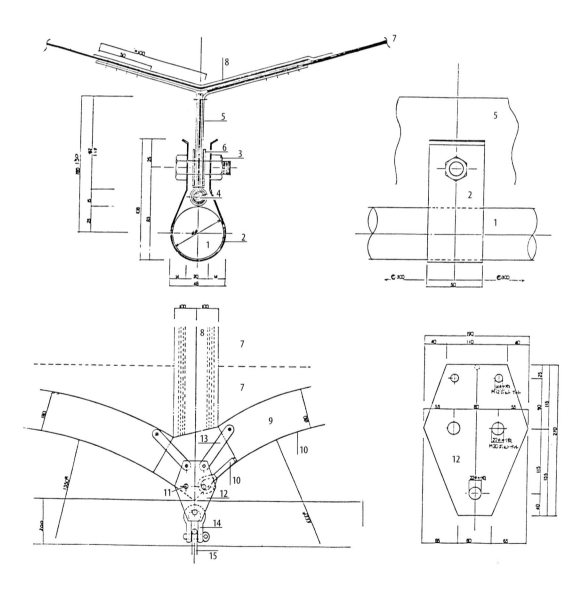

View from access road

Roof of Government Stadium, Hong Kong

Design

A lack of international-standard sports facilities have long been a problem for Hong Kong; almost six million inhabitants on 1000 km^2 inevitably leads to other priorities in land use. The solution, as often in Hong Kong, came from the Royal Hong Kong Jockey Club, which operates the only legal betting franchises in the Territory. It commissioned proposals for a modernisation of the Government Stadium in So Kon Po, south of Causeway Bay on Hong Kong Island. Four schemes were investigated, two renovation schemes and two renewals. Each of these preliminary designs were analysed on a cost-benefit basis taking into account design, construction, marketing and financial factors.

Brief | 40 000 seats, 75 % covered by a roof
- Stadium designed for football and rugby, but without an athletics track
- Provision of 50 corporate boxes, each with its own catering facilities
- Facilities for major pop concerts
- No underground car park
- Construction scheduled so that the annual Rugby Sevens Tournament could take place
- Construction period not exceeding 3 years

Tendering | In the tender documents the construction period was set at 36 months. For the structure in-situ concrete frames and slabs with precast concrete bleachers were anticipated, for the roof two main arches spanning 240 m north to south with a rise of 55 m. Steel trusses at 6 m centres were to span between the arches and the upper edge of the precast concrete rakers. Dragages et Travaux Publics also offered an alternative design (beside their regular offer) with a construction period of only 24 months and a considerable cost discount based on a modified structural concept:
- The secondary beams and floor slabs of the concourses were precast concrete elements, and
- the steel trusses of the roof came at 12 m centres instead of 6 m.

Their offer was conditional on the design team being under the direction of the contractor and not the architect, so that the working drawings could be prepared in a sequence and without communication problems. After assessment of the offer the contract was awarded to Dragages.

Client
Royal Hong Kong Jockey Club

Architects
Hellmuth, Obata & Kassabaum, Inc., Sports Facilities Group

Structural Engineering
General, membrane, structure and substructure: Ove Arup & Partner Hong Kong Ltd.
Detail design steelwork: Bond, James, Norrie, Marsden

Specialist consultants
Wind tunnel tests: Rowan, Williams, Davies & Irwin Inc, Guelph, Ontario
Geotechnics and traffic engineering: Ove Arup & Partner Hong Kong Ltd.
Quantity surveyors: Davis Langdon & Seah
Resident engineering for structural and building services work: Ove Arup & Partner Hong Kong Ltd.

Contractors
General contractors: Dragages et Travaux Publics
Subcontractors of the membrane: Permafab Pty Limited

Completion
1994

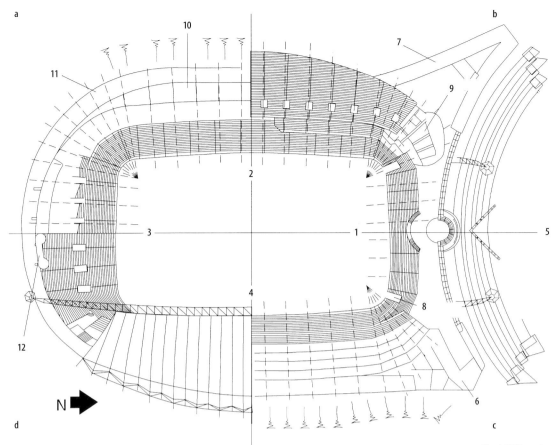

Plan 1:2000: a main concourse, b upper seating, c suite level, d roof; 1 north stand, 2 west stand, 3 south stand, 4 east stand, 5 main entrance, entrance from the Eastern Hospital Road, 6 east ramp, 7 west ramp, 8 pitch access, 9 restaurant, 10 toilets and food outlets, 11 loop road, 12 screen

Development of the roof design | The project was subject to decisive external constraints and an extremely tight time schedule. The roof form was determined largely by the topography of the steep So Kon Po valley, also the shape and massing of the stadium was supposed to reflect its situation in the valley.

In the attempt to have the structure follow the roof edge, almost by chance the concept of a wide-span truss arch emerged alongside other variants (a curved space-frame roof was also explored). This expressive form became further emphasised by enlarging the arch span to the total length of the stadium. Thereby the arches developed into a major design element at the entrance. For architectural reasons the arch arrangement had to be symmetrical.

A rock slope in the valley, on the far side of the loop road, offered a meaningful abutment position. The arch then had to rise steeply to provide vehicle headroom, and becoming cycloid in a plane 12° to the vertical. Trusses span from the arch to concrete rakers at the top of the stands. The roots of the arch were detailed as pins, which is more an architectural feature than a structural necessity.

Structure

The main arches are two-pin trusses with a square cross-section (3.5 m x 3.5 m), which are inclined at 12° to the vertical. The chord members are steel tubes (Ø 406 x 12.5 mm, Ø 406 x 16 mm and Ø 406 x 20 mm); as membrane connection a halved square hollow section (Ø 400 x 400 x 12.5 mm) is welded on. The length of an arch truss module is approx. 5 m; they have been manufactured in pairs, so that one arch is made up of 24 pairs of straight units. The web members, verticals, horizontals and diagonals are CHS (Ø 193 mm to 273 mm).

The curved secondary roof trusses are hip frames pin-supported at the raker ends. They span between 40 m and 55 m and have a three-chord section 3.5 m deep. They are connected to the arch by stainless steel bearing pins. At the support on the rakers a vertical truss frame element acts as a column.

Bird's-eye view

View of roof

View of arch abutment

View of the reinforced concrete seating frame

The valley cables (Ø 80 mm), midway between the secondary beams, are prestressed against the horizontal truss members and a simple truss element in the upper plane of the arch. These horizontal trusses transfer their compression loads into the secondary beams; the horizontal components of the valley cable forces are thus balanced at the roof plane.

Load Assumptions | The value of the wind forces was finally to be confirmed in wind tunnel tests. These were carried out with a standard typhoon windspeed as described in the HK Code. An aeroelastic wind tunnel model was built to explore the interaction of the roof structure with the wind. The resulting wind loads were higher than expected, the critical load case was the northerly wind direction parallel to the arch.

This made it necessary to reinforce the structure and the membrane and also resulted in an increase of the design loads for the glazing.

Bracing | Along the upper edge of the stands the structure is stiffened through cross bracing. The vertical elements of the roof trusses are connected by truss members and thus stabilise the secondary trusses. They carry the wind loads in the longitudinal direction of the roof and transfer them into the stand structure.

The roof girders are connected with each other by structural elements (Ø 273 x 16 mm, Ø 273 x 20 mm, Ø 219 x 12.5 mm) proceeding parallel to the arch and are thus braced, to provide lateral stability and to resist the horizontal wind forces from the roof membrane. These structural elements are anchored in the last roof girders, equipped with an additional cross bracing (Ø 355 x 16 mm) between the bottom chord members, whereby the horizontal forces are transferred to both sides, into the arch and into the cross bracing along the top edge of the stands.

Membrane | The covered plan area of the roof is a gigantic 8000 m^2. To minimise the self-weight of the structure, a PTFE-coated glass fibre fabric (in the end a Shearfill® membrane supplied by Chemfab) was proposed as a roof membrane. This kind of covering was new for Hong Kong in this scale, but was considered appropriate by the authorities and the client.

On both sides the roof was subdivided into five large panels, covering three roof bays each. The membrane is a flat saddle shape between the arch upper chords and the valley cable. Next to the arch it is nearly flat, increasing its curvature towards the outside edge. The membrane sag relative to the valley cable corresponds to the depth of the roof girder.

The valley cables together with the edge clamps serve for the prestressing of the membrane. Under wind load they transfer the major part of the wind uplift.

Along the edges the membranes are connected by a series of clamping plate strips, while they are not fixed at the upper chords. Along the roof girder they are clamped with aluminium U-sections, boltropes and Neoprene padding onto an aluminium base plate, serving as construction aid. The plate in turn is bolted onto a cold-worked sheet metal section, welded onto the upper chord of the roof girder. Along the edges at the main arch, which carry less load, a similar detail is used, where the U-section is replaced by a clamping strip made from aluminium flat.

In the end bays the membrane edge is formed by an edge cable, thus avoiding bending of the edge girder. There the membrane is clamped with clamping plate strips and boltrope and joined by stainless steel U-straps (Fl 50 x 5, a = 400 mm) with the edge cable (Ø 80 mm).

A splash plate is fastened parallel to the clamping plate strip as a shallow gutter.

Construction

First year | Demolition of the old stadium began the day after the Rugby Sevens Tournament of 1992. Only the old pavilion was retained to house the changing rooms for the 1993 games. In the first year the entire podium structure was erected where in the following summer, 1993, the Rugby Tournament took place. Construction resumed after the event.

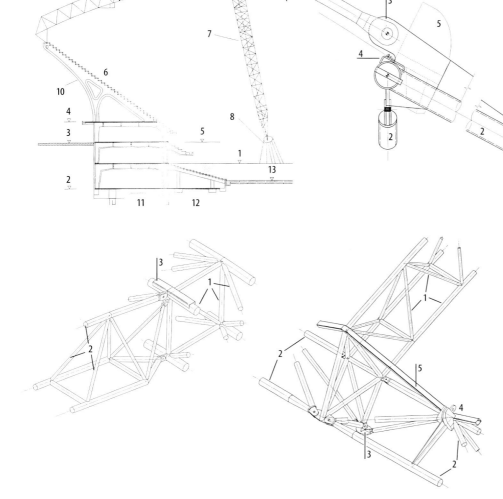

Section 1:1000: 1 main (entrance) concourse, 2 service level, 3 old access road, 4 upper concourse, 5 suite level, 6 upper stand, 7 steel truss arch, 8 reinforced concrete abutment, 9 roof girders (steel truss), 10 reinforced concrete frames, 11 caissons, 12 reinforced concrete pad foundations, 13 pitch

Edge detail at upper edge of stands with valley cable connection: 1 valley cable with fork fitting, 2 bracing, 3 connecting plate, 4 membrane connection (clamp), 5 sheet metal gutter

Structural isometric, roof girder connection at the main arch: 1 main arch, 2 roof girder, 3 halved square hollow section welded on for membrane connection

Structural isometric, roof girder connection at upper edge of stands: 1 roof girder, 2 truss bracing, 3 column base at concrete support, 4 valley cable connection, 5 rectangular tube for membrane connection (clamp)

Second year | The main tasks of the second year were to erect the roof, complete services and finishes and lay the permanent pitch along the edges. Roof girders and arch sections were assembled on trestles on the pitch and welded up from components, before the last layers of corrosion protection were applied. Each arch was assembled by crane masts in four sections supported at the joints, and simultaneously a working platform was provided, from where the in-situ splice welds were made.

After the final coat of paint for the steelwork the roof membrane was assembled.

The panels were delivered from Australia and mounted on a scaffolding above the rear bracing. Light cables were attached to the free edge and the membrane pulled towards the arch by tirfors fastened at the truss chord. The loose, deployed fabric was gradually hauled to its connection points via threaded bars connected at the edge clamping strips. After all the clamped edges were fixed, the valley cables were lifted up from the pitch, the rear pin installed and the pre-tension applied by a jack on the arch.

Following erection of the roof, work could start on the construction of the permanent pitch and installation of the seating.

View

Convention Centre, San Diego, California, USA

Design

At the seaport of San Diego, the architect chose a membrane roof reminiscent of sails as a meaningful and appropriate landmark. It covers 9000 m^2 of exhibition area on the otherwise open uppermost floor of the new exhibition and conference centre. The four-storey building, with main exhibition and conference rooms of 23 000 m^2 on the ground floor, is situated at the edge of the city centre directly on San Diego Bay.

The main structure was erected in conventional construction, and large, sculptured, triangular buttresses support the membrane roof 10 to 30 m above forming a dominant architectural feature. Originally planned as riveted steel girders with allusions to historic harbour structures, they had to be executed in reinforced concrete due to problems of fire protection and building costs. The architectural concept was developed in the office of the architects Arthur Ericson & Partner; the roof concept and the structural planning originates from Horst Berger & Partner consulted later in the design process. By the time Berger & Partner was called in the general structural concept was already worked out: a membrane roof supported by flying struts suspended between external concrete buttresses.

Roof structure

The membrane roof consists of six bays, which span unsupported over the entire width between the concrete buttresses (91.5 m). The inside bays are 18.3 m wide, which corresponds to the distance between the buttresses; they are displaced, however, by half a bay width, so that the valley cables are in line with the buttresses while the flying struts and ridge cables are positioned on the centre line between them. Both end bays are held by edge cables and overshoot the last buttresses by 9.15 m.

In the main membrane, eye-shaped ventilation openings are placed at the centre of each bay between the high point struts. Above them a continuous secondary membrane acting as rain protection is suspended from the tips of the flying struts of the main roof. Its low points, which are tied down to the middle of the valley cables of the main roof, serve at the same time as drainage onto the main membrane, where the rainwater runs off along the valley cables.

The flying struts and the main supporting cables holding them as mentioned lie on the centre line between two buttresses. These double cables penetrate through the roof skin, bifurcate and proceed to the top to be joined to the two adjacent buttresses. The ridge cables are arranged over the main supporting cables; they are joined with the flying struts at their upper

Client
Port of San Diego, Unified Port District

Architects
General: Arthur Ericson & Partner
Membrane structure: Horst Berger & Partner
Management and construction documents: Deems Lewis McKinley
Programming and facilities planning: Loschky, Marquart & Nesholm

Structural engineering
Buildings: John A. Martin & Associates and Johnson, Spurier
Membrane and structure: Horst Berger & Partner
HVAC, installations: Syska & Hennessey
Foundations: Woodward-Clyde Consultants

Contractor
General contractor: Tuto-Saliba-Perini
Membrane, steel and erection: Birdair Inc.

Completion
1990

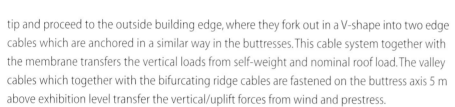

Roof structure, computer perspective
Building cross section

tip and proceed to the outside building edge, where they fork out in a V-shape into two edge cables which are anchored in a similar way in the buttresses. This cable system together with the membrane transfers the vertical loads from self-weight and nominal roof load. The valley cables which together with the bifurcating ridge cables are fastened on the buttress axis 5 m above exhibition level transfer the vertical/uplift forces from wind and prestress.

The mast tips are joined with each other and with the lower anchor points to the buttresses through a system of safety cables lying above the membrane. These cables hold the masts in case of a catastrophic failure of the membrane, serve as assembly aids and hold the masts in position, before the membrane is installed or in case of a later replacement of the membrane. In addition they stabilise the structural system and prevent a slackening of the structural cables under extreme wind suction.

At the open ends, where a guy against the buttresses or onto another edge structure was not possible, a horizontal compression strut is used, extending over the entire length of the roof. It has forked ends to carry the horizontal forces in longitudinal direction of the building and to balance them over the entire building length. The edge cables are anchored here at the fork ends. An additional vertical guy is not available either. The long compression strut is suspended from the tips of the flying struts via short hanger cables and guyed for stability onto their lower end.

Roof; interior view

Roof membrane with structure and compression truss with fork ends

Upper buttress connection

Load | The design wind speed is 31 m/s (112 km/h); the wind loads were determined in a wind tunnel test. The roof also designed for a vertical roof load of 12 lb./sq.ft. (0.58 kN/m^2), required according to the building code.

Steel | For the steel structure the following material grades were used: rolled steel sections and steel flat are structural steel ASTM A36, tubes are ASTM A53. The high-strength bolts used correspond to ASTM A325. The flying struts consist of steel tubes (CHS Ø 457.2 x 9.5 mm, l = 18.2 m). The horizontal compression truss, extending over the entire length of the roof has a square cross section (a = 0.9 m, with chord tubes Ø 168.3 x 7.1 mm and diagonals Ø 48.26 x 3.68 mm) and forked ends (chord Ø 141.3 x 6.55 mm and diagonals Ø 42.16 x 3.56 mm). All tension members (main structural cable Ø 47.6 and Ø 54 mm, ridge cables Ø 28.6 and Ø 19 mm, valley cables Ø 50.8 and Ø 28.6 mm) are galvanised spiral strands to ASTM A586 and are equipped with a PVC-sheath as additional corrosion protection due to the highly corrosive atmosphere by the sea.

Membrane | The roof membrane consists of Shearfill®, the PTFE coated-glass fibre fabric by Chemfab. Along the sides in front of the buttresses the membrane is connected with the Y-branches of the valley cables. Due to the large forces very flat edge cables result with a horizontal span of 9 m. The membrane is joined with the edge cable via a clamped edge of aluminium strips and by aluminium straps positioned at regular intervals. Along the edges a cold-worked aluminium sheet (t = 9.5 mm) is clamped onto the roof membrane and serves as a gutter.

At the top of the flying struts, underneath the fixed connection of the external security cables, the membrane and ridge cables are fixed on a movable prestressing element, guided along the mast tube. Before the membrane is connected, it separates from the ridge cable and is clamped to a separately prestressable steel ring and connected there via a boltrope.

The cables meeting at the membrane corners are fastened through fork fittings, in part adjustably, to a welded steel plate assembly, anchored in the concrete buttresses by anchor bolts. The membrane there has circular cut-outs reinforced with a fabric patch. Due to the large membrane forces a lightweight corner cable is provided, which is connected onto the large edge cables with light cable fittings. A movement of the membrane parallel to the edge cables is prevented through a swaged cable clamp.

Roof plan 1:1000

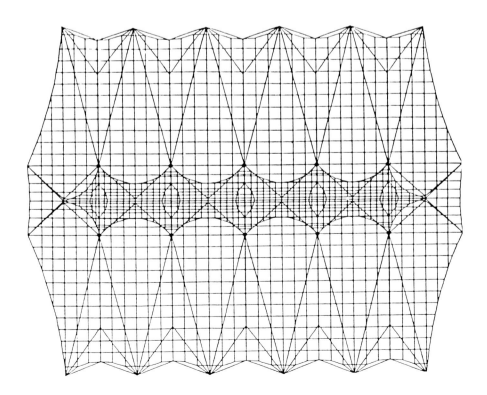

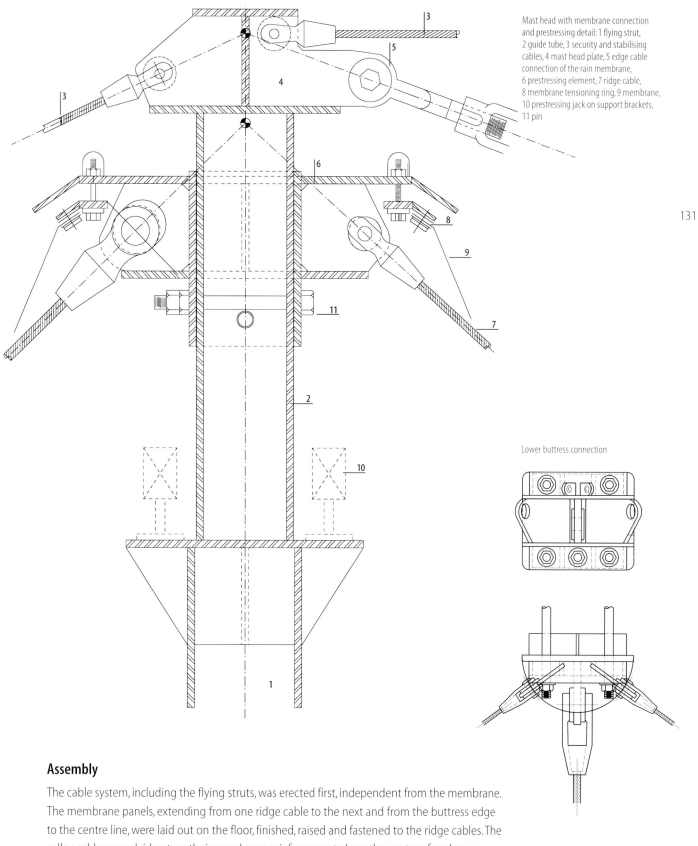

Mast head with membrane connection and prestressing detail: 1 flying strut, 2 guide tube, 3 security and stabilising cables, 4 mast head plate, 5 edge cable connection of the rain membrane, 6 prestressing element, 7 ridge cable, 8 membrane tensioning ring, 9 membrane, 10 prestressing jack on support brackets, 11 pin

Lower buttress connection

Assembly

The cable system, including the flying struts, was erected first, independent from the membrane. The membrane panels, extending from one ridge cable to the next and from the buttress edge to the centre line, were laid out on the floor, finished, raised and fastened to the ridge cables. The valley cables were laid out on their membrane reinforcement along the centre of each membrane element and fixed in position through a short lacing. The prestressing of the main membrane is accomplished by jacks, set onto welded support brackets; they jack the movable mast head up which is then secured by a pin.

View with parking lot

Roof Canopy of Sainsbury's Supermarket, Plymouth, Great Britain

Design

The eleven curved canopies are reminiscent of sails, forming a line in front of Sainsbury's supermarket and supplying the conventional shed with a both impressive and functional accent. They are one of the last works of Peter Rice, who died in 1992, and were finished by Martin Manning from Ove Arup & Partners.

Original concept | The search for the optimal technical solution was surely not completed in this project as Peter Rice had intended it. He wanted to have many different components working together in the system. His original design shows many-branched tree columns of steel tubes and a curved primary steel grid, filled-in and braced by prestressed tension ties, a wooden lattice with a weather skin above, made from a coated man-made fibre fabric. The strength and load-bearing capacity of the components varies according to the scale of the grid. The different grid sizes make a many-layered geometry, and the different structural components work together as space structure.

In the tender version the structure consisted of one tree column per sail with 12 and/or 8 steel branches and a primary grid (Ø 114 mm) with square meshes measuring 5 x 5 m in plan, braced by a system of prestressed stainless steel ties in the mesh corners. A wooden secondary grid of rectangular laths 70 x 25 mm with a mesh width of 60 cm was placed on it. It was covered with a prestressed membrane, which was to be raised in the convex area by distance slats

Client
J. Sainsbury plc

Architects
General and membrane structure:
Jeremy Dixon, Edward Jones

Structural engineering
Membrane, structure and substructure: Ove Arup & Partners, Peter Rice, Martin Manning

Contractors
General contractors: Birse Construction
Membrane: Landrell Fabric Engineering Ltd.
Steel : Westbury Tubular Structures Ltd.

Completion
1994

Model photograph: Tender version with secondary timber grid

Elevation view

and in the concave area was to be anchored linearly by aluminium boltrope sections. Membrane, primary and secondary grid as well as the cable bracing together made up a structural surface with a structurally usable shear stiffness in the plane of the grid.

Executed version | In their built form the roofs have lost something of their original innovation and gained in economy, which may also have been the result of the client finding the project slightly too extravagant. The basic principle survived; the structure, however, was simplified: The tree columns were reduced to three 'branches' and a tie bracing (not executed), the secondary grid was omitted, and instead the steel tube grid was refined to 7 x 10 meshes, the tube diameter was increased and the roof membrane was conceived as a minimal surface in the shape of a soap film, so as to be able to prestress the roof membrane directly across the grid without the need for intermediate supports. The less-layered structural hierarchy was felt to be an acceptable compromise. The increased simplicity also has a certain desirable formal effect.

Rear view

Situation

Altogether 11 sails of identical shape (each 20 x 10 m, with a maximum height of 16 m), overlapping each other, constitute a 175-m long roof canopy along a parking lot. In the east it abuts to a newly planted row of trees. Sails and trees work together, to improve the local wind regime

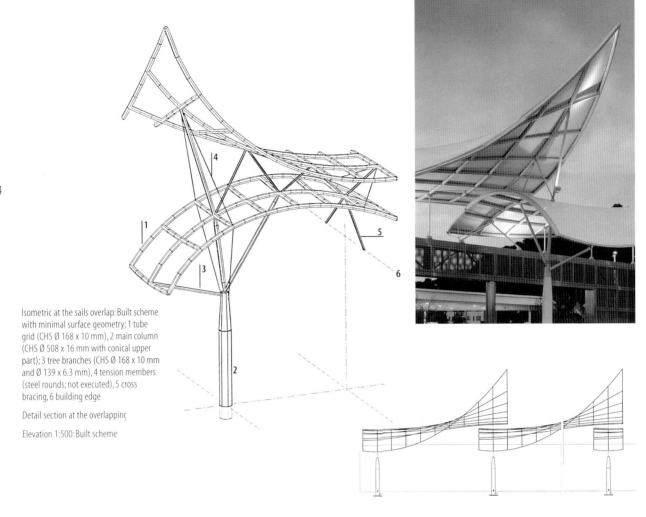

Isometric at the sails overlap: Built scheme with minimal surface geometry; 1 tube grid (CHS Ø 168 x 10 mm), 2 main column (CHS Ø 508 x 16 mm with conical upper part); 3 tree branches (CHS Ø 168 x 10 mm and Ø 139 x 6.3 mm), 4 tension members (steel rounds; not executed), 5 cross bracing, 6 building edge

Detail section at the overlapping

Elevation 1:500: Built scheme

in the parking lot and in front of the supermarket. The prevailing wind direction is from southwest, from the sea and the estuary of the river Plym, to which the supermarket building turns its back. The sails redirect the fast-moving air up even higher and provide a wind-sheltered zone over the parking lot. They are more like wind deflectors than sails. These and other special arrangement like a 'rain fence' and a porous grid wall were explored in wind tunnel tests at Bristol University.

Structure

This is not a membrane structure in the true sense, but this in no way diminishes its architectural charm. The membrane here is rather a cladding for a tubular steel grid, shaped according to the form-giving laws of physics governing the shape of prestressed membranes; it lies on the grid and is anchored there against uplift.

Tubular grids orthogonal in plan form a doubly-curved surface, the form of which makes it act as a space structure. It is point-supported on three sides – at two opposite sides by the tree columns, each with three steel-tube 'branches' and a cable bracing. Along the building it is supported by vertical struts and cross bracing arranged on the axis of the tree column. This main column consists of steel tubes Ø 508 x 16 mm with a tapering, conical upper part; the tree branches are CHS Ø 168 x 10 mm and Ø 139 x 6.3 mm. Each sail consists of a steel tube grid of 7 x 10 meshes with a tube diameter of Ø 168 x 10 mm. In the overlapping area the upper sail is raised above the lower one by tube compression members (Ø 139 x 6.3 mm) proceeding diagonally from node to node.

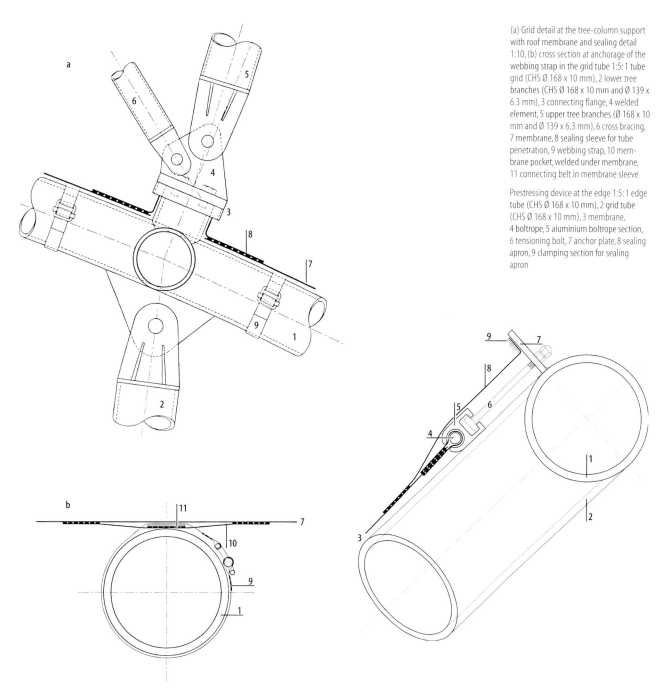

(a) Grid detail at the tree-column support with roof membrane and sealing detail 1:10, (b) cross section at anchorage of the webbing strap in the grid tube 1:5: 1 tube grid (CHS Ø 168 x 10 mm), 2 lower tree branches (CHS Ø 168 x 10 mm and Ø 139 x 6.3 mm), 3 connecting flange, 4 welded element, 5 upper tree branches (Ø 168 x 10 mm and Ø 139 x 6.3 mm), 6 cross bracing, 7 membrane, 8 sealing sleeve for tube penetration, 9 webbing strap, 10 membrane pocket, welded under membrane, 11 connecting belt in membrane sleeve

Prestressing device at the edge 1:5: 1 edge tube (CHS Ø 168 x 10 mm), 2 grid tube (CHS Ø 168 x 10 mm), 3 membrane, 4 boltrope, 5 aluminium boltrope section, 6 tensioning bolt, 7 anchor plate, 8 sealing apron, 9 clamping section for sealing apron

The continuous edge tubes and the ones in transverse direction were bent into a three-dimensional curve; the other mesh direction is welded in between them. Each grid was prefabricated as elements in the steel yard and site-welded together on the ground by means of special welding templates.

The sails are braced by diagonals along the building front, by cantilever columns together with a triangular bracing in the plane of the tree branches and by the in-plane shear stiffness of the grid.

Membrane | The roof skin consists of PVC-coated polyester fabric. Each sail was manufactured in one piece; the seams follow the grid lines. The membrane is prestressed with 2 kN/m against the edge members of the grid.

Along the straight edges the membrane is anchored through a boltrope running in an aluminium boltrope section, prestressed via threaded bolts and closed by a welded-on weather apron with self-tapping screws clamped to a steel flat, running along the edge.

After prestressing the membrane is sealed at the points where the support struts penetrate the upper sail by a glued fabric sleeve and strapped to the tubes with webbing straps and buckles.

Oblique view

Service Station, Wanlin, Belgium

Client
FINA Europe sa, Brussels

Architects
General and membrane structure:
Samyn et Associés spl, Brussels
G. André, Y. Avoiron, L. Finet C. Geldof,
D. Mélotte, Ph. Samyn

Structural engineering
General, structure and substructure:
SETESCO, G. Clantin, P. Samyn
Membrane structural analysis:
Vrije Universiteit Brussel, afdeling architectuur
M. Mollaert

Contractors
Membrane and steel: De Boer Tenten nv, Bree with Technet Gmbh, Berlin (cutting patterns)
Concrete, foundations: Gobiet sa, Seraing
Interior finishes: F. Moureau sa, Ans

Completion
1995

For the 75th anniversary of its founding PETROFINA wanted to erect an extraordinary, forward-looking service station. Analogous to the increased performance and comfort of automobiles these new service stations in their style have come a long way from the traditional image of a steel canopy with a colourful advertising frieze in gaudy colours along the roof edge.

The site lies on the E 411 between Namur and Luxembourg on both sides of the motorway and will be connected by a bridge-type restaurant to be erected by the managing company. Other areas contain the usual functions of children's playground, picnic area and car and truck parking. Open terrain, fields and woods in a hilly landscape typify the site, which is exposed to the local harsh weather without protection.

Design

The design omits loud advertising; only the 'totem pole' and the company logo in front of the sales area show the company´s name. The large glazed windows of the shop offer a view onto the grandiose landscape. In the interior of the shop a large room height provides ample space and good air quality. The large air volume reduces the need for frequent air renewal. And the overall style is quite unlike similar projects: there are no varnished or coated surfaces, but instead galvanised steel, coated fabric, anodised aluminium, wood, glass and concrete.

Each of the two stations consists of a continuous translucent textile roof, 2000 m^2 large, which covers the whole plant including the petrol station, reception area and sales buildings, and protects customers and visitors from sun, rain and snow. The standard landscaping, common along the motorways, of hedges and lawns is replaced by shallow, reflecting ponds, which have a balancing effect on the local microclimate. The shop and sales area was executed as a large glazed box (h = 5.4 m).

Structure

The structure under the membrane consists of 3 parts:
– the roof girders,
– the inner and outer compression struts, and
– the tension ties joining roof girders and compression struts.
The steel structure is galvanised.

The main structural elements are curved truss girders, two with a length of 40 m and one 44 m long, with a rise of 4.6 and 7.0 m. They form the substructure for the membrane roof. These

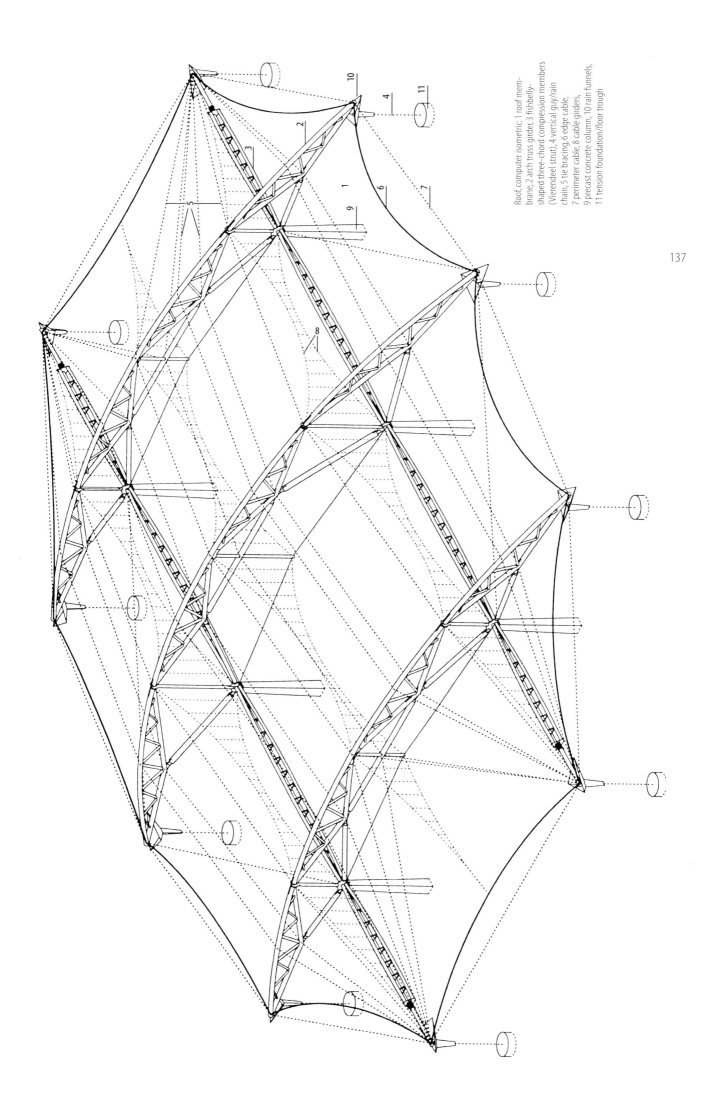

Roof, computer isometric: 1 roof membrane, 2 arch truss girder, 3 fishbelly-shaped three-chord compression members (Vierendeel strut), 4 vertical guy/rain chain, 5 tie bracing, 6 edge cable, 7 perimeter cable, 8 cable girders, 9 precast concrete column, 10 rain funnels, 11 tension foundation/floor trough

Side view

Partial view from the rear

arch girders consist of partly straight, partly curved steel tubes (Ø 114.3 x 3.6 mm to Ø 219, 1 x 8 mm). The roof skin is connected to the upper chords (Ø 193.7 x 5.4 mm) with adjustable threaded U-bolts and welded onto connection plates (a = 1.0 m).

Each girder is supported on two 5.5 m-high concrete columns on a 20 x 15 m grid. In longitudinal direction of the roof, perpendicular to the arch girder, the column heads are connected by horizontal, three-chord tubular steel compression members with a fishbelly shape (3 x Ø 133 x 5 mm). In the end bays these compression struts cantilever out by 12 m and are, in the same way as the cantilevering arch ends, tied down and anchored by vertical tension ties.

A system of round bar ties (Ø 20 to Ø 36 mm) joins the arch girders to brace them perpendicular to their plane against lateral buckling and asymmetrical wind or snow loads from the membrane. These tension ties are fastened to the ends of the cantilevering compression struts and also carry their self-weight.

Due to the horizontal forces acting on the girders their upper chord members are loaded by tension and the bottom chord partially by compression, which shows in the member cross sections. Perimeter cables proceeding in chord direction of the cable edge hold and stabilise the cantilevering compression struts in a horizontal direction and carry forces in a tangential direction.

Curved tension cables under the membrane are part of a lightweight cable girder which stabilises the membrane and stiffens it against wind loads. The connection between membrane and structure is covered by 252 transparent PVC shells.

The steel weight of this structure, relative to the covered plan area, is 14.5 kg/m^2 and relatively highly, compared with other textile roofs of this kind. On the one hand this results from the flat membrane curvature, whereby relatively large forces emerge under prestress and loads. On the other hand it is due to the fact that the roof is not guyed against the foundations, but spans freely from column to column, so that the horizontal forces are transferred by the structure in plane of the roof where they are balanced.

The design snow load is 0,45 kN/m^2. A vertical wind suction load of 1.0 kN/m^2 was assumed.

Membrane | The fundamental decisions on the architectural form were made with the help of simple physical models from cardboard, nylon stocking material and thread, which could be built within a few hours. Rough order-of-magnitude calculations for the forces were later verified through a computer analysis, which was more time-consuming.

The structure consists of four independent membrane elements, two inner and two edge panels. After establishing the basic geometry a computer calculation yielded excessive displacements of the membrane under extreme wind load. For the remedial stiffening of the structure three different schemes were explored:

reinforcing the membrane by cables, inserted in pockets, which was considered a rather expensive solution;

enlarging the rise by increasing the arch curvature (i.e. smaller radius);

Interior view with reinforced concrete column, cable girders and horizontal, three-chord tubular steel compression members (3 x Ø 133 x 5 mm) with fishbelly shape

View from shop

installing a system of light cable girders underneath the membrane.

The cable girder system was the preferred solution, which also left the girder geometry unchanged. The vertical cables of these girders are slightly prestressed; they do not carry a share of the snow loads.

The cable girders lie partly on the axis of columns and compression struts and partly between them, in the latter case their horizontal forces are introduced into the edge cables.

Under prestress the membrane stresses are 16 and 11 kN/m in weft and warp direction; under load they are 21 and 19 kN/m.

For reasons of availability the used membrane is stronger than statically necessary. It consists of white PVC-coated polyester fabric with an acrylic coating on both sides; it has a weight of 1450 g/m^2 and a tensile strength of 8/7.8 kN/5 cm in warp/weft direction (FERRARI Précontraint 1502/144, required would have been only a 1202/144 with a strength of 5.7/5.2 kN/5 cm). The total factor of safety of the membrane (including the reduction factors) lies at 5.0.

The coated fabric is classified as 'difficult to ignite' (class M2 in the French code). It is manufactured in a strip width of 1.8 m, the high frequency welds are 4 cm wide. The stainless steel edge cables run in cable sleeves.

Through careful pattern cutting and production, wrinkling of the membrane was almost totally avoided. The compensation factors for cutting the pattern were determined as 1.10 % in warp and 1.15 % in weft direction from stress-strain diagrams from physical testing by the fabric manufacturers.

Edge, membrane corners | The stainless steel edge cables (Ø 12 to 30 mm) run in membrane sleeves and are continuous along the arch girder over the entire length. Corner plates are clamped onto the upper and lower side of the membrane. These plates are cut with a radius along the edge cables and connect the membrane via adjustable, threaded U-bolts and welded connection plates (a = approx. 1.0 m) with the curved upper chords (Ø 193.7 x 5.4 mm) of the trusses.

At the ends of the compression struts two edge cables and up to ten tension ties are connected. The edge cables are pin-joined with a threaded bolt, running in a tube sleeve welded to the corner assembly at the end of the compression strut.

Arch covering, drainage | Overlapping, transparent PVC panels along the ridges of the steel arch girder cover the open garland edge of the membrane and at the same time serve as natural ventilation. Their fire performance is in the same class as the membrane. The very cost-effective PVC-material used is not fully transparent, but it is sufficient to make the steel-girder structure behind it visible.

The rainwater is collected in simple membrane gutters welded on along the edge cables, stiffened along their edge with thick boltrope edges and held by sheet metal angles. The water is channelled into large, triangular sheet-metal funnels, hanging at the ends of the truss girders

Rain funnel at compression member with guy cable: 1 membrane, 2 compression strut (end tube of three-chord compression member), 3 corner plate with U-bolt, 4 edge cable with eyelet fitting, 5 adjustable bolt with fork head, 6 tube sleeve, welded onto 7 welded head plate assembly, 8 membrane gutter with thick boltrope edge and brackets, 9 sheet metal rain funnel, 10 vertical guy cable with fork fitting, 11 funnel connection

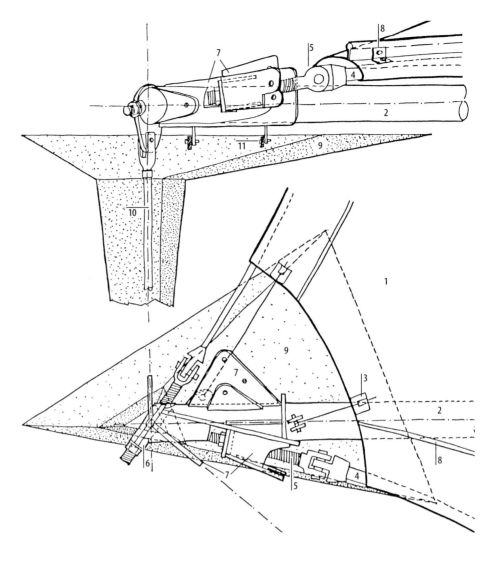

Membrane connection at tubular truss: 1 chord tube of truss girder, 2 roof membrane, 3 edge cable sleeve/HF-weld, 4 garland edge cable, 5 corner plate, 6 threaded U-bolt with domed cap nut, 7 connection plate, 8 bolt with cotter pin

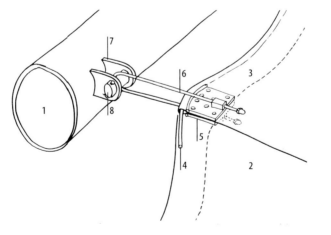

and at the guys, guided down along vertical rain chains (no downpipe) and collected at the ground in concrete troughs. The membrane gutters at the edge also prevent snow sliding off in winter.

Canopy roof A (in the foreground), link B and pedestrian roof C

Airport, Salzburg, Austria

Increased traffic at Salzburg airport in recent years made it necessary, and economically possible, to carry out a reorganisation and redesign of the 'landside', i.e. the side connected with ground traffic, to make it more user-friendly and to improve the traffic system. The principal features of the improvement scheme were a six-storey, 300-m long car park and the reorganisation of the airport forecourt with a generous roof, covering four lanes and smaller roofs over the footpaths between the car park and the service building.

Requirements

The roofs had particular requirements and therefore demanded special solutions: As the forecourt of the airport, with its roads, parking areas and footpaths is used rather heavily, restraints emerged for the desired roof constructions and technical requirements, such as differing clearance heights for car and pedestrian traffic and safety distances, which had to be observed. The goal of the architects was to develop a roof shape, which, in addition to its function as weather protection, would provide some visual orientation on the forecourt and generate a light, friendly and positive atmosphere. The present concept was named first choice in the limited competition because of its dynamic membrane shape and a form language corresponding to the movement of flying. It was later chosen for execution. The project consists of four principal parts:

1. A large canopy (roof A, 110 m x 25 m) as roofing over the drop-off area in front of the entrance of the existing service building.

2. A covered pedestrian zone (roof C, 350 m x 2.5 m), along the parking garage.

3. The 'links' between canopy roof and pedestrian zone, consisting of three parts (roof B, one 18 m x 7 m and two 45 m x 7 m).

4. A 'logo structure' for the parking garage, consisting of three sails mimicking the nearby Alpine mountains.

The following wish list refers to the desired properties of the roof cover:

– A cover which is translucent but avoids glare in strong sunlight to provide shading, without making it too dark;

– Dirt on the surface should not be visible from below;

– Easy maintenance and cleaning and good durability;

– Good resistance to hail and snow and noiseless when it rains;

Client
Salzburger Flughafenbetriebsgesellschaft m.b.H., Salzburg

Architects
General: Michael Rhomberg, Grossgmain
Membrane structures: Horst Dürr, IF Ingenieurgemeinschaft Flächentragwerke, Constance-Reichenau

Structural engineering
Membrane structures: Horst Dürr, IF Ingenieurgemeinschaft Flächentragwerke, Constance-Reichenau
Steel structure and foundations: Dipl.Ing. R. Herbrich, Salzburg

Contractor
Membrane: Carl Nolte GmbH & Co, Greven
Steel: Ferroglas, Hörsching

Completion
1994

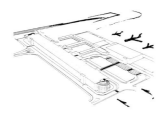

Perspective view with forecourt and new car park (in front), not showing, however, the membrane roofs

– A manageable technology in detail, with a high degree of prefabrication and a short assembly period;
– Cost effective and economical construction.

Design

All four structures exhibit a dynamic steel and membrane form and careful detailing.

The canopy roof A swings up, opens towards the car park and draws the visitors into the service building.

The pedestrian zone roof C moves in waves along the parking garage.

The large undulations of links B join the car park ('the ground') with the service area ('the air/flight'). With their dynamic shape they seem to 'fly' over the road with ease at the required clearance height of 5 m.

The logo, as the highest point of the roofscape, serves as goal and guidepost for the travellers and visitors arriving by car, and conducts them into the car park. The design represents three snow-capped peaks.

The roofs are pure rain shelters and have no additional function.

Structure and membrane forms | All four structures are mechanically tensioned membranes with different geometries:

The canopy roof (roof A) consists of linearly added arched surfaces and corresponds to the type of the arch arcade; it consists of saddle surfaces, rectangular in plan, formed by parallel arches and intermediate straight edges. The links (roof B) and the pedestrian zone (roof C) are wave-shaped ridge and valley forms (wave surfaces).

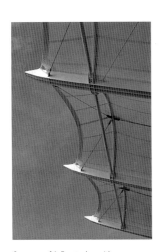

Canopy roof A: Front edge with membrane, arch and tension tie

Substructure | As a counterpart to the tensile structure, the works above the substructure are primarily in compression. As a compression system some unusually curved and formed steel tube structures (pedestrian zone, roof B and C) are used, and four mast trestles carrying steel truss girders (canopy roof A) suspended from them. The link structures (roof B) have similarity with the spars of a ship.

These three structures have no guys or tension foundations, as the membranes are braced by steel frames at roof level. The steel compression structure transfers the forces to the ground via compression pad foundations. The logo is the only tensile membrane guyed via steel masts and compression struts.

Canopy roof A | The primary structure consists of four mast trestles of circular tubes, with four triangular truss girders in two rows suspended from them. These carry the 24 main girders of the membrane roof, curving slightly upwards and perforated with circular web openings. Their bottom chord consists of a horizontal H section, which simultaneously serves as gutter, as anchor point for the membrane and as prestressing support for the transverse arch girders. The arch girders are made from circular tubes with a horizontal steel tube tie separately connected with a pin. The arches are held laterally by a stabilising cable running in the longitudinal direction of the membrane and anchored in the end bays at the longitudinal girders by means of two bifurcating cables.

Links (roof B) | Roof B consists of two similar parts of different lengths. A slightly upward curving hollow box girder, with square cross section is supported by four twin columns of circular tube, which together with the connecting elements form a gable frame. The columns are fixed at reinforced concrete foundations. Cantilevering from the hollow box section to both sides are tapering members carrying on the outside a curved wave-shaped edge girder made from a circular tube, on which the roof membrane is fixed.

Pedestrian zone (roof C) | Roof C covers the pedestrian area along the car park and consists of a continuous, undulating membrane surface of rectangular elements in plan, with a wave-shaped

Canopy roof A: Plan with steel structure and with membrane roof 1:1000

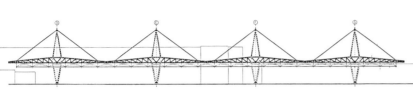
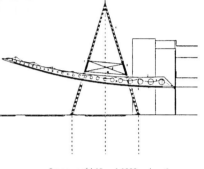

Canopy roof A: View 1:1000 and section 1:500

Canopy roof A: Plan view with masts and triangular truss girders

Logo D

front and a straight rear edge. Tube columns encastréed in the foundations carry curved cantilevers with a raised gutter from a cold-worked C-section, into which the membrane is prestressed. The front edge is formed by a steel tube in which the membrane is anchored.

Logo (part D) | Three four-point sails each forming an upright triangle represent the peaks of the Salzburg Alps. For architectural reasons they consist of planar membrane surfaces, which are not in fact sensible structural elements at this size, because technically they could not be realised without extreme difficulties. Therefore they had to be modified into an anticlastic (saddle-shaped) surface by a ridge cable lying in the membrane surface and stabilising the sails. They are anchored at their top by a short tubular mast and two guys connected in the roof of the car park's stair tower. On the ground they are anchored in four reinforced concrete foundations and the walls of the building.

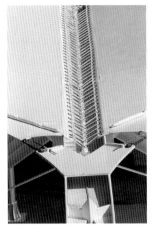

Link B: Lacing along the central gutter beam

Roof membrane connections

The roof membrane of all four buildings consists of PVC-coated polyester fabric type II with a PVDF topcoat.

Canopy roof A | At the bottom chord of the transverse girder the membrane is connected by welded-on circular steel keys and short tube sections inserted into a membrane sleeve. For the prestressing of the membrane the arch girders are jacked up from a jacking seat and fastened adjustably by bolts to steel plates welded onto the main girder. Along the building the roof membrane is held by a straight, rigid element – a gutter section similar to the connection along the transverse girder – and at the front edge of the roof by an edge cable.

Links (roof B) | Along the edges the membrane is connected by welded-on circular steel keys and short tube sections inserted into a membrane sleeve. Where necessary, a membrane gutter proceeds along the membrane edge clamped onto fixing brackets. To avoid a prestressing detail with a complicated geometry, the membrane was fastened and tensioned by a lacing parallel to the central girder: over the central girder a welded steel sheet valley gutter is placed, into which the membrane is laced around a circular tube lying in the gutter and fastened to welded-on steel plates. To enable the lacing, the membrane edge is equipped with a boltrope and a row of eyelets. At the narrow sides the roof skin is held through edge cables. On the outside edge either a 'rain boltrope' is fixed, i.e. a round foam rubber piece welded into a membrane sleeve, or a membrane gutter is clamped onto the snow grid.

Canopy roof A 1:15: Transverse girder (1) with membrane details; the bottom chord (2) of the transverse girder is made from an I-section, in which the membrane (5) is held by welded circular steel keys (6) and short tube pieces inserted into membrane sleeves. In transverse direction curved arch girders (3) are arranged, made from circular tubes and equipped with a tube tension tie (4) connected separately via a circular bolt. To prestress the roof membrane the arch girders are jacked up from a jacking seat (7) with bolt holes for seat fastened by bolts (8) with an adjustable connection at the transverse girder by means of welded-on steel plates.

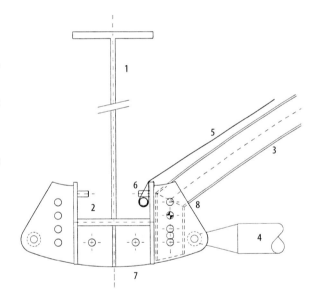

Link B, membrane details 1:10:
Left: Membrane edge with membrane gutter clamped to bracket; 1 edge tube, 2 membrane, 3 circular tube inserted into membrane sleeve with steel keys, 4 gutter brackets (snow grid), 5 membrane gutter, 6 clamping section
Right: Inside gutter (5) with lacing (3) and eyelet edge (2) onto a circular tube (4) connected at the gutter.

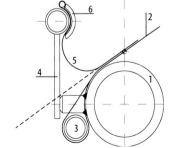
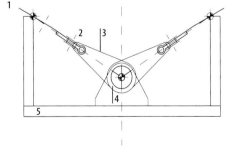

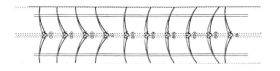
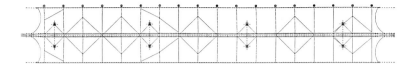

Link B: Plan, longitudinal and transverse sections 1:500

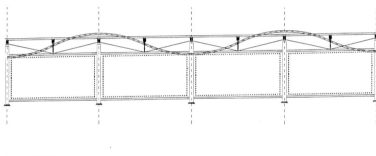

Pedestrian zone C: Typical plan, longitudinal and transverse sections 1:200 (the areas between the columns serve as advertising spaces)

Pedestrian zone C

Pedestrian zone (roof C) | Above the tube cantilever a sheet metal prestressing gutter is connected, with prestressing bolts welded on. Two adjacent membrane panels, furnished with boltropes, are held with steel-flat clamping strips and prestressed into the gutter below with nuts.

The membrane connection along the front edge is similar to the one along the sides of roof B, i.e. short tube pieces in a membrane sleeve. At the back edge the membrane is led over a horizontal CHS and anchored behind by rectangular tubes inserted into membrane sleeves and held by a prestressing bolt arrangement; around these bolts the membrane is cut out in a circular shape.

Logo (part D) | At the membrane corners bent triangular corner plates with a central opening join the oncoming ridge, edge and guy cables with each other. The edge cables run in membrane sleeves reinforced with a webbing. They cannot be prestressed separately, but are connected directly at the corner plates through fork fittings. Valley cables and guys have a swaged, threaded fitting. The webbing strips are connected adjustably through light fork turnbuckles.

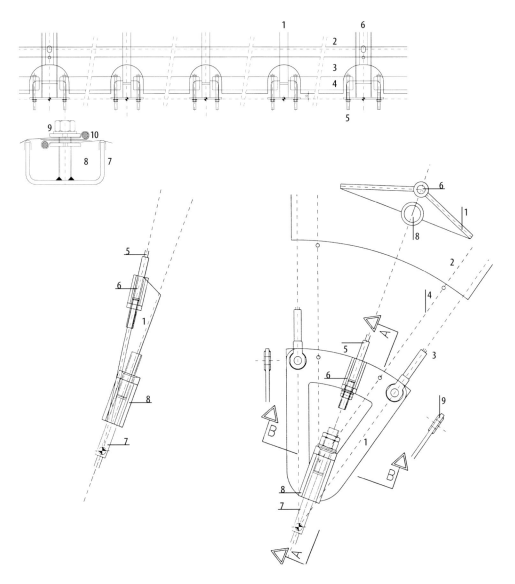

Pedestrian zone C

Above: Rear membrane anchorage with rectangular tubes inserted in sleeves and prestressing bolt arrangement, 1:25; 1 arch girder, 2 membrane, 3 edge sleeve, 4 rectangular tube, 5 spanning device with threaded bolt, 6 prestressing gutter/membrane joint

Below: Detail of prestressing gutter, 1:5; sheet metal gutter (7), welded-on prestressing bolt (8), two membrane panels with edge boltrope (10) and clamping plate strips from steel flat (9)

Logo D, details 1:10: Triangular bent corner plate (1) with (5) ridge cable, (3) edge cable and (7) anchorage of guy cable. Valley and guy cables have a swaged threaded fitting, the edge cables are equipped with fork fittings and connected to the corner plates. The webbing connection is not shown.
1 corner plate, 2 membrane, 3 edge cable with fork fitting, 4 webbing connection (not shown), 5 ridge cable with swaged threaded fitting, 6 tube sleeve, 7 guy cable with swaged thread fitting, 8 tube sleeve, 9 plate reinforcement/shimming washer

View from the Avenida de Maria Cristina with V-shaped edge support

Auditorium Roof CAMP DE MART, Tarragona, Spain

This roof over an open-air theatre with 3000 seats situated at the edge of the old city centre of Tarragona has some special features, which distinguish it from others of its kind. It lies underneath the city walls and below a monumental tower in a natural bowl forming the amphitheatre at the foot of a steep embankment. Tower and city wall serve as backdrop for the open-air theatre, whose membrane roof with its fascinating form has become the landmark of this part of town.

Design

The allusion of a manta ray from the sea, of which the roof form is reminiscent, emphasises the connection of the city with the Mediterranean. The view of the passing access road, the Avenida de Maria Cristina, and onto the tower and the ancient city wall remain unchanged.

The axis of symmetry of the amphitheatre and of the structural axis coincide and lead onto the monumental tower. The masts of the structure rise up high and lean toward the tower. In a common frame of reference the tip of the V-shaped main mast becomes the visor point, redirecting the view onto the tower. The structure appears to be tied in with measure and proportion between membrane roof and tower. Seen from inside the natural backdrop is emphasised by the steel tubes acting as a picture frame. At the same time the translucent roof creates a light-filled space without strong shadows, giving a feeling of openness, of being in the open.

Situation

The auditorium with its semi-circular plan has the form of an amphitheatre with steeply rising seats. The entrance is from above, through a circular access path on the level of the access road in front of the theatre. The stage is hexagonal; on its left and right stage houses are erected in conventional reinforced concrete construction.

Structure

The main structure, nearly semi-circular in plan, consists of a roof membrane, strengthened at the edges by steel cables. It is supported by four V-shaped mast trestles around the edge, and by one A-shaped main mast on the axis of symmetry behind the stage. It is anchored at five guy cables between the lower V-masts, fastened to ground anchors as well as by one large guy cable on the axis of symmetry holding the main mast.

Despite the flat membrane shape there is just sufficient surface curvature available at each

Client
Ayuntamiento de Tarragona

Architects
General: Ayuntamiento de Tarragona; Josep Ollé I Ribé, Angel Martínez I Lanzas
Membrane structure: Studio Tensoforma

Structural Engineering
Membrane, structure and substructure: Studio Tensoforma

Specialist consultants
Foundations, geotechnics: Lorenzo Jurina, Politecnico di Milano
Lighting: Ayuntamiento de Tarragona; Julio Cadena Gutierrez

Contractors
Membrane and steel: Tensoforma Trading s.r.l.
Substructure, concrete & foundations: Rodio s.a.

Completion
1993

Plan view of roof with contour lines, 1:1000

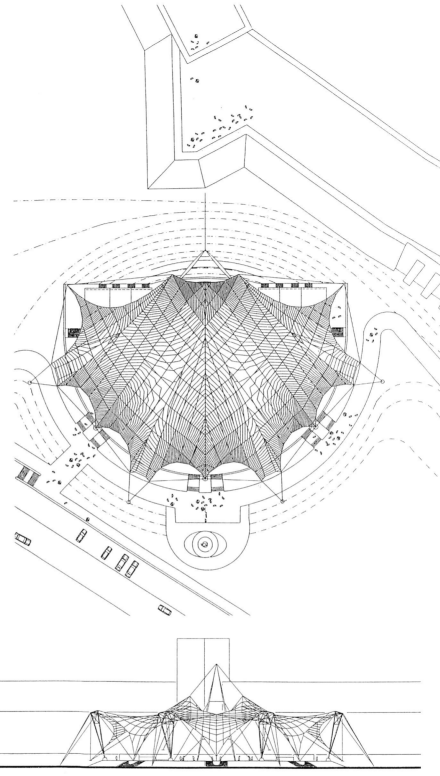

Views in direction of the structural axis and perpendicular to it, 1:1000

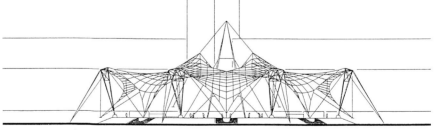

Opposite page:

Side view from above; edge mast with space frame head assembly and main mast

Corner detail of guy structure, edge mast with space frame head assembly and anchorage of the security cable with light tube mast: A security cable spanning across the roof prevents collapse of the main mast in case of catastrophic failure of the membrane

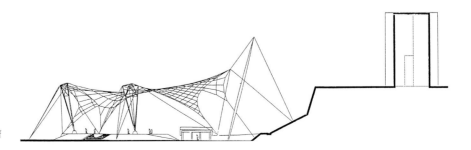

point. To avoid large displacements under snow load a system of snow cables was introduced in the middle of the roof.

A continuous ring cable joins the high and low points; its function is to hold the V-masts laterally, which are only guyed by three cables converging onto one base point.

The roof drains naturally to the low points into floor inlets there. A gutter at the membrane edge is not provided; when rainfall is light, the clamping section along the edge acts as rain deflector, when rain is heavier, the water can overshoot the edge.

The structure was analysed for the load cases of dead weight, snow and wind. An evenly distributed snow load of 40 kg/m^2 and a wind load of 110 kg/m^2 (corresponding to a maximum wind speed of 151 km/h) were assumed.

Masts | The large A-mast as a structural system is a tripod (h = 38.4 m) with two intermediate guys. Two pin-supported masts joined at the top, with a dumbbell-shaped cross section (h_{max}=1.2 m), consisting of two steel tubes (2 Ø 323.9 x 8.35 mm) connected by a twin web with variable height, are tied to the outside by a main stay on the axis of symmetry and also by two additional guy cables on both sides connected at intermediate points along their length. They consist of three modules and one head piece; these modules are joined by bolted flange connections for easier handling during assembly. Where the intermediate guys are connected to the masts a horizontal tie cross-bracing consisting of steel rods (Ø 24 mm) joins the mast tubes. The mast base is formed by a steel ball joint (Ø 320 mm).

The head of the main mast consists of a welded steel plate assembly and circular tubes. The guy cables and other ties are either connected with fork fittings and conical sockets at the head plates or they pass through a tube sheath where they are anchored with nuts and counter nuts. To simplify transport the mast head assembly is joined with a bolted flange connection to the mast members.

The heads of the smaller, V-shaped edge masts (h = 16 m, tube Ø 353.9 x 6.3/7.1 mm) were developed into a space frame consisting of three tetrahedrons (3.2 m x 6.0 m, h = 4.4 m). It consists of steel tube compression members (Ø 168.3 x 4.5 mm, Ø 159 x 3/4.5 mm, Ø 108 x 3/4 mm) and horizontal round steel tension ties (Ø 45 mm) and serves as a fixing point for two membrane tips and a ridge cable between them. To avoid bending of the mast (i.e. due to the difficulty of carrying bending moments across the spherical nodes of the space truss) two additional cables (outside Ø 26 mm, inside Ø 36 mm) were introduced, leading to the mast base. To the outside the edge masts are guyed by three cables (Ø 36 mm + 2 x Ø 19 mm).

The head structure of the smaller, V-shaped edge masts is connected to the lower mast tubes by a spherical space frame node. For this purpose the mast tube cross section is reduced in diameter by a welded steel cone, held at the node by a welded tubular sleeve and secured by bolts. The lower compression members of the space frame are attached to conically turned steel parts secured by bolts. The turned steel parts are fastened at the node by bolts coaxial with the member.

At the upper nodes of the head structure, which is a welded assembly of steel plates and tubes, the upper space frame struts are connected via steel plates and pins. The horizontal tension ties are either connected with fork heads and bolts or are furnished with threads, passing through a welded tube sheath and being secured there with nuts and counter nuts.

The four inner compression tubes at the head of the smaller edge masts are connected with the welded nodes by a bolted flange with a diameter equal to that of the connected tubes.

The detailing of the bases of main and edge masts is similar: Two converging tube sections penetrate each other at the base and are either welded to a transition piece made from steel plates (main masts) or connected with a bolted flange (V-shaped edge masts). The transition piece transfers the mast forces to a spherical steel bearing, whose bottom shell is fastened on the r.c. foundation with anchor bolts. A cover plate fastened with bolts onto the bottom part protects the bearing.

The steel parts of the main structural elements are galvanised and painted.

View of the tower with A-shaped main mast and guy cables

Ropes, guy cables | The guy cables are connected adjustably with a conical socket with female thread and with two threaded extension bars at a solid round steel transom flattened above and below, which is fastened to a welded universal joint made from two crossing bearing pins.

The tension ties are galvanised steel cables with cable fittings and prestressing devices with the following dimensions:
- hanger cables at the main mast Ø 52 mm,
- guy cables at the main mast Ø 52 mm + 4 x 2 Ø 26 mm,
- membrane corner guys Ø 52 mm, Ø 42 mm, Ø 26 mm,
- edge cables Ø 42 mm to Ø 20 mm,
- snow cables in membrane surface Ø 64 mm to Ø 36 mm,
- ring cables Ø 36 mm to Ø 19 mm.

Guy point at membrane corner with valley cable

The large edge cables, furnished with fork fittings with a built-in universal joint, are anchored together with a transverse bearing plate with turnbuckles and with the valley or ridge cable with its end socket in a welded steel corner element.

The version of the corner plate assembly without valley or ridge cable is much simpler: The edge cables are connected with fork fittings and the single transverse bearing plate with two threaded bars at a welded steel corner plate, which in turn is anchored by two tension ties and a bearing pin to a steel element fastened to the concrete foundation.

Membrane | The roof shape can best be described as a modified symmetrical high point membrane. Along the edge there are five high points with different heights, the highest one lies on the axis of symmetry. At five intermediate low points the membrane is tied down. To avoid large heights and long membrane tips the high points were 'modified', i.e. dissolved into two points with a short edge cable lying between them. The high points are either anchored in V-shaped twin columns or connected by two tension ties to the tip of the A-shaped mast with an intermediate guy between them proceeding down to the ground.

The membrane consists of a high-strength PVC-coated polyester fabric (Ferrari Precontraint-1502-Fluotop®) with a tensile strength of 8/7.8 kN/5 cm (warp/weft). The material is classified as 'difficult to ignite' (class 2 to Italian, class M2 to French standard). It is coated with Fluotop®, a 5-µ strong PVDF-polymer coating thus extending its service life. The coating forms anti-adhesive protection film against smog and other air pollutants and so visually prolongs the life span of the membrane.

For the membrane and their textile additions, edge cable pockets, membrane aprons etc., a reduction and safety factor was required between four and five to guarantee a very long service life. (The safety factor applied for coated fabrics on the short-term strength determined in load tests is made up of a real safety factor, allowing for a variation of the manufacturing quality and the applied loads as for conventional building materials, steel, concrete, and one reduction factor, which allows for the decrease of the fabric strength due to long-term loading, ageing and effects from handling, transport and erection. By selecting a higher reduction factor a longer design life of the membrane may be achieved.)

Along the edge the web membrane is held and reinforced by clamping strips. Due to the large membrane forces – the membrane strength corresponds approximately to a type V – and because of the corresponding size of the edge cables (max. Ø 42 mm) and the associated fittings the edge was executed as a clamped edge with a steel boltrope. The clamping section is intermittently fastened to the edge cable by sheet metal straps in short spacing. The boltrope has a threaded fitting at its end and is anchored in a tube sleeve welded to a transverse steel plate onto which the membrane is also connected at the corner by means of sheet metal straps.

Foundations

Due to the flat curve and the high loads the supports and guys of the tensile structure suffer relatively high tension or compression forces. A geological survey showed very unhomogeneous soil conditions, so that the foundation design was adjusted according to the particular situation. For tension piles and compression foundations the same pile type was used, namely

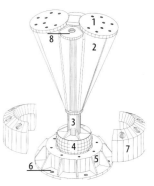

Edge mast during assembly: Steel tube compression member, spherical node and turned cones to attach the compression struts of the space frame. In the background the installation of the main mast using a scaffold substructure on the slope can be seen

Isometric: Mast base detail at V-edge mast: 1 tube flange connection of the mast tubes, 2 welded transition piece, 3 round steel, 4 steel spherical bearing, 5 bearing bottom, 6 bore for anchor bolts, 7 covering, 8 cable anchorage

steel micropiles (injection piles of the type Ropress/Tubfix). According to ground type and load, up to 16-m long injection piles were used, for the compression piles to reach a load-bearing stratum, free of mud inclusions or for the tension piles to activate a sufficiently large earth volume to be used as ballast weight.

Assembly

The steel structure was assembled and erected within 16 days, for installation and prestressing of the membrane a further 25 days were required.

The installation proved extremely difficult due to the problematic terrain conditions at the rear edge of the amphitheatre, where a mobile crane could only be used with difficulty. Only a mobile crane with a capacity of 15 t could be used, the A-shaped main mast however weighs 20 t and is positioned in an inaccessible location underneath the wall. It had to be divided into parts, assembled on the construction site and pulled up the sloping terrain and into its final position with tirfors. Scaffolding had to be erected on the slope and a chute, on which the mast was dragged with tirfors horizontally to its foundations, without loading the crane with the full mast weight.

Interior view

Climate-Control Parasols for the Extension to the Prophet's Holy Mosque, Medina, Saudi Arabia

Design

The commission to design a convertible shade roof for two large courtyards of the Holy Mosque of the Prophet in Medina presented an unusual architectural and technical challenge. At this location, where tens of thousands of believers gather every day, the climatic conditions had to be improved without destroying the character of the open quadrangles and their familiar environment.

The solution consisted of twelve convertible parasols, 17 m x 18 m large and with a height of 14 m at the eaves in their opened condition, which fit in perfectly with the proportions of the courtyards. With a span of 24 m, measured diagonally across the corners, they are the largest ones built of this kind.

The six parasols with their funnel-shaped membranes create the effect of translucent vaults, spanning between the columns and the arched arcades surrounding the courts, and produce a large free space. Their timeless form with its carefully designed ornaments match harmoniously with the traditional architecture.

Technical installations

Air conditioning | Through controlled use of this variable roof structure extreme climate differences can be equalised, modifying the climate inside the building and reducing energy consumption. In summer the opened parasols offer shade during the day, when the white PTFE-membrane re-emits a large proportion of the solar radiation to the outside. At night, in a closed position, they allow the thermal energy stored in the heated building surfaces of the courts to escape unhindered into the night air.

In winter, when the median temperatures are relatively low, the parasols are opened during the night to prevent an extreme cooling of the building; they are closed during the day to admit the mild winter sun and so to warm the building's surfaces and masses.

Opening and closing times of the parasols are computed electronically for every day according to the position of the sun, dependent on the seasons, and allowing for weather conditions, outside temperature, wind and cloud cover and the results fed into the parasol controls. To make the climatic conditions in the courtyards more comfortable during the summer months, when the air temperatures in the shade can exceed 45 °C, the parasols' controls are coupled with the building's air-conditioning system. Air outlets in the base and the capital of the parasol

Client
SBG Saudi Binladin Group, Jeddah

Architects
General: Dr Kamal Ismail
Membrane structure: Architekturbüro Dr Bodo Rasch

Structural engineering
Design and structure: SL Sonderkonstruktionen und Leichtbau GmbH
Structural analysis: Buro Happold

Specialist consultants
HVAC: Fa. Krantz, Dr-Ing. Haaf
Project management, quantity surveyors: Buro Happold + SL Sonderkonstruktionen und Leichtbau GmbH

Contractors
General contractor: Saudi Binladin Group
Subcontractor: SL Sonderkonstruktionen und Leichtbau GmbH
Membrane: KOIT High-Tex GmbH
Mechanics and steelwork: Liebherr-Werk Ehingen GmbH

Completion
1992

Section through quadrangle and mosque, 1:100

154

Plan 1:500: 1 courtyards with parasol roof, 2 the Holy Mosque of the Prophet, 3 extension, 4 movable dome structure

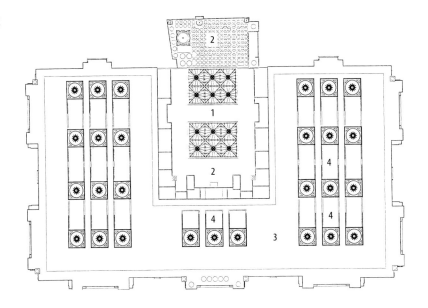

Aerial view of the courtyards with parasols during assembly

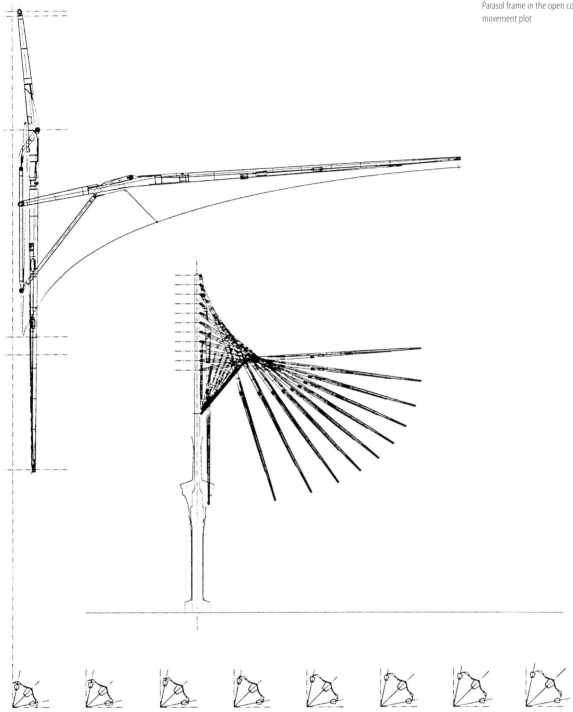

Parasol frame in the open condition and movement plot

Parasol arms and head with gold-coloured finial, arms, struts, protection flaps and light hanger cables for the support of the membrane during movement and folding

Erection

Plan of parasol

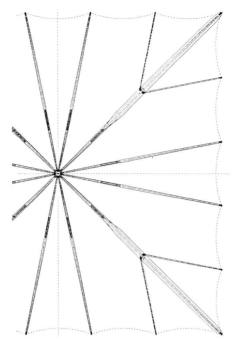

column distribute cool air noiselessly in a wide area, so that the entire quadrangle is cooled evenly and effectively.

Drainage | When it rains, the rainwater is drained via the funnel shaped membrane and through the column into the floor drain.

Lighting | Each parasol has four strong lights installed in the cladding above the column capital, which illuminate the quadrangles at night.

Structure | The components of the parasol structure, mast column, arms and struts, are a welded construction of a high-strength fine-grain steel.

For good precision of the movements the bearing bores of the critical hinges were made by a (computer-controlled) NC-milling machine. This manufacturing precision of the movable parts together with the placement of the hydraulics pumping in a central plant room makes an almost noiseless working of the parasols possible.

Foundations | The reinforced concrete pad foundations of the parasols are joined with each other by ground beams.

Mechanics | The parasol has altogether 20 arms: four long diagonal arms with two short passive arms each connected to them, eight middle arms, of which four are shorter and four longer.

The parasol is opened and closed by a hydraulic cylinder on the column axis, whose upper end is pin-connected with all the active arms. In the closed condition the hydraulic cylinder has driven out completely at the top; to open the parasol it drives down, whereby the struts, which are pinned above the capital, push the parasol arms, which are connected at the upper end of the hydraulic cylinder, to the outside, prestressing the parasol membrane. The electronic controls and hydraulic pump are housed in a central controls room in the basement and connected with the 12 parasols via high pressure pipes under the marble floor of the courtyards.

Protection flaps | In designing the parasols special care and consideration was given to the closed condition and to the folding of the membrane. Pin-connected protection flaps of especially lightweight carbon-fibre-reinforced resin laminate are fixed along the diagonal arms.

They are moved by a mechanism of struts and hinges installed in the arm and driven by the arm movement. They close around the membrane, when the parasols are closed. When the parasols are closed these protection flaps, together with the fixed sheet-metal flashing in the upper part of the arms, form a stiff envelope for the light fabric membrane.

Loads | Wind tunnel tests for the specific situation in the courtyards served as a basis for the design of the structural elements of the parasols, which were designed for the opened and closed condition using a wind speed of 155 km/h. An anemometer connected with the central controls prevents the opening and closing at wind speeds above 10 m/sec (36 km/h).

Membrane | The funnel-shaped membrane is made from white PTFE-fabric. Edges and ridges are reinforced by a webbing. The underside of the fabric is decorated with ornaments in blue colour. The fabric was developed specially for this project, its thread thickness and type of weave were modified repeatedly, until all structural and mechanical requirements on strength and durability were fulfilled. The material is resistant against UV-radiation, chemicals and fire, has a low surface friction and is therefore a near-ideal material for convertible sun roofs.

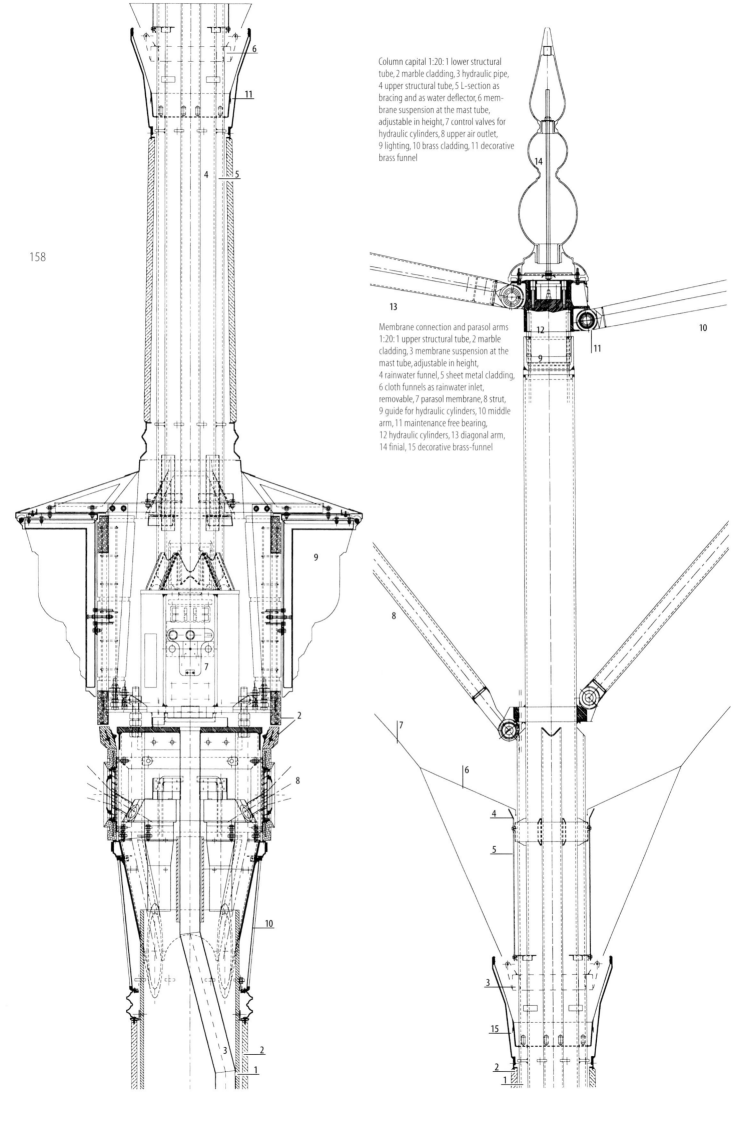

Column capital 1:20: 1 lower structural tube, 2 marble cladding, 3 hydraulic pipe, 4 upper structural tube, 5 L-section as bracing and as water deflector, 6 membrane suspension at the mast tube, adjustable in height, 7 control valves for hydraulic cylinders, 8 upper air outlet, 9 lighting, 10 brass cladding, 11 decorative brass funnel

Membrane connection and parasol arms 1:20: 1 upper structural tube, 2 marble cladding, 3 membrane suspension at the mast tube, adjustable in height, 4 rainwater funnel, 5 sheet metal cladding, 6 cloth funnels as rainwater inlet, removable, 7 parasol membrane, 8 strut, 9 guide for hydraulic cylinders, 10 middle arm, 11 maintenance free bearing, 12 hydraulic cylinders, 13 diagonal arm, 14 finial, 15 decorative brass-funnel

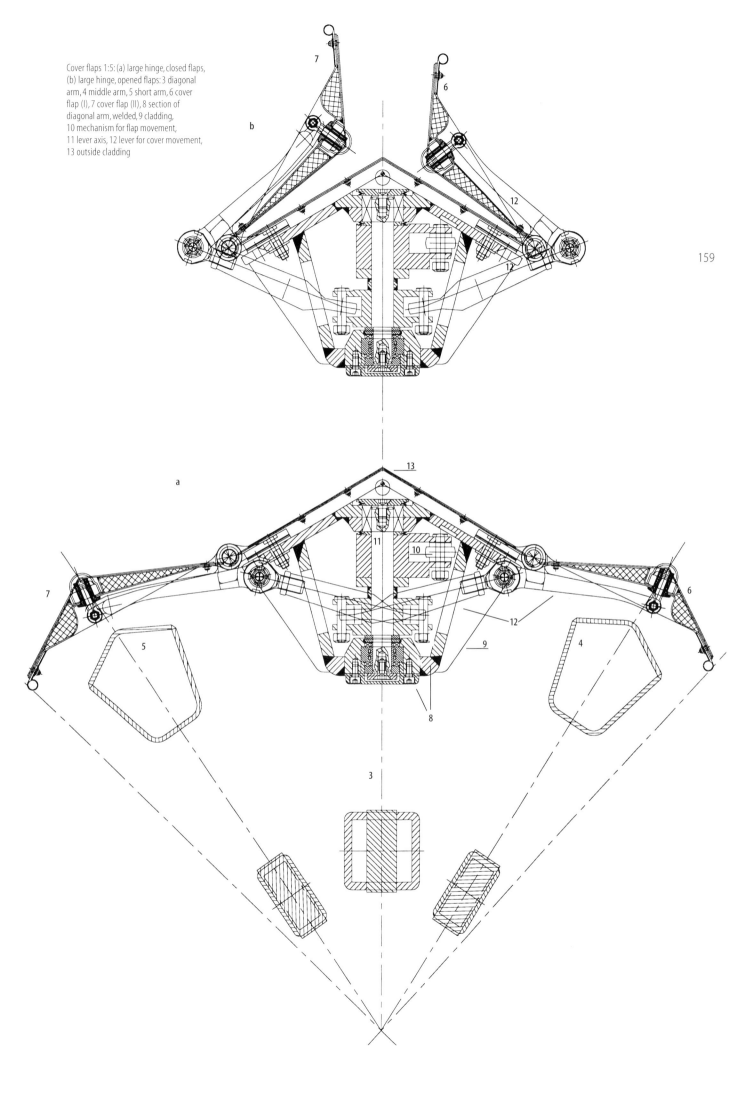

Cover flaps 1:5: (a) large hinge, closed flaps, (b) large hinge, opened flaps: 3 diagonal arm, 4 middle arm, 5 short arm, 6 cover flap (I), 7 cover flap (II), 8 section of diagonal arm, welded, 9 cladding, 10 mechanism for flap movement, 11 lever axis, 12 lever for cover movement, 13 outside cladding

Square module 10 x 10 m, additive arrangement of 4 x 8 modules, with partial membrane wall in the arch plane

Mobile Building System for Exhibitions "AIC-Tensoforma"

Client
Tensoforma Trading, Bergamo

Architects
General and membrane structure:
Architekten und Ingenieure Coop,
Stuttgart (Design: Jürgen Hennicke)
and
Studio Tensoforma, Bergamo
(Architect: Stefano Bertino)

Structural engineering
Membrane and structure: AIC
Architekten und Ingenieure Coop,
Stuttgart

Contractor
Membrane and steel: Tensoforma
Trading, Bergamo

Completion
1987 and following years (different applications)

Membrane structures, including tents, usually have guys, which carry the tensile forces from the roof membrane down into the ground; anyone who has ever stumbled over tent chords and pegs on a campground will know this. If a modular, additive building system is desired, the task is especially troublesome. The membrane building system described here manages without external guys.

Development

In the seventies the office of AIC in Stuttgart developed a membrane building system with four braced arches carrying a roof membrane over a square plan. This system was taken over under license by Studio Tensoforma, an Italian planning office for membrane structures, and developed into an exhibition building system in co-operation with the AIC. It offers an installation floor, inside drainage as well as planar and curved space enclosing membrane walls. Marketed by Tensoforma Trading, it was used successfully on numerous sites in Italy, France, Russia and Spain. As required for a building system, all essential aspects of storage, transport and erection were optimised in a long-time development and in the course of various project applications.

Design

Aesthetics and style were of great importance in the design. The architects see their modular arch structures as a modern analogy of Romanesque cross vaults. However that may be the simple, classical lines and consistent technical detailing of these structures fit into almost any surroundings.

At the beginning there was the self-imposed task of developing a building system, which was mobile, i.e. capable of being erected and dismantled with simple means, and which was modular, i.e. capable of being added to in two directions. Above it the membrane roof was to be square in plan (10 x 10 m). Later it was developed into a modular system with square, hexagonal and triangular modules.

Form | The system includes three basic modules, square, hexagonal and triangular, which can be set up alone or in combination. Numerous combinations of the three basic modules are possible, as the edge arches have equal dimensions. The versions for short-term and permanent applications have the same standard dimensions.

The square module has a saddle-shaped roof membrane over four circular edge arches with

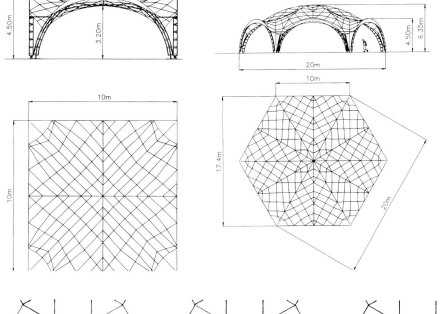

Square module 10 x 10 m, 1:500

Hexagonal module 10 x 10 m, 1:500

Modular construction for hexagonal, square and triangular modules 1:1000

valley cables diagonal in plan. Geometrically the form can also be interpreted as a section of a vertical square prism with two HP surfaces penetrating each other at a right angle. Four semi-circular arches in this case replace the parabola-shaped edge arches of the HP surface.

The triangular module is set up similarly. The form of the hexagonal module is a composite surface of six saddle shapes between vertical edge arches and three radial inner arches.

Structure

Depending on the type of application the four semi-circulars edge arches are three-chord trusses with curved round steel diagonals or a three-chord Vierendeel arch with gusset plates, the latter being simple to make, but it is stressed in bending and therefore heavier. The arches are connected by bolts at the apex and at the quarter points and pin-connected at the arch base with a steel-base plate, where also the diagonal valley cables are anchored.

The temporary version is anchored at the ground with earth nails and the permanent version with rawlplugs or resin anchors in concrete foundations or a ground slab.

The temporary (summer) version is only designed for wind load; while the ones for long-term application are designed for wind and snow loads.

Square module | The square modules available in four sizes (6 x 6, 8 x 8, 10 x 10 and 12 x 12 m) consist of a roof skin of PVC-coated polyester fabric over four vertical steel edge arches with two diagonal valley cables as membrane reinforcement. The arches are connected with each other by four diagonal struts, fixed at 2/3 of the arch height, so that the stability of the arch is not dependent of the membrane.

In the plane of the arch the structure works as a two-pin arch, perpendicular to it as a space structure with bending and torsion. The building is braced in the plane of the arch by the arch itself, and perpendicularly to it by tubular members in tension or compression connected at half-height across the corner between adjacent arches. Two valley cables diagonal in plan transfer the wind uplift forces directly to the base-plates, without loading the arch.

Hexagonal module | The elements of the hexagonal module have an edge length of 10 m; each module covers a base area of 260 m^2. Three radial and six edge arches carry the membrane. The edge arches are joined with each other by six diagonal struts, whereby the stability of the arch is maintained without use of the membrane.

Hexagonal module 10 x 10 m, with inner arches, interior view

Square module 10 x 10 m, additive arrangement on steel tube column bundles, interior view, arch constructed as Vierendeel truss

Square module 10 x 10 m, additive arrangement on reinforced concrete columns, interior view, arch constructed as truss arch with round steel diagonals

For ease of transport and erection each middle arch consists of six, each edge arch of four elements with three-chord section. The elements are connected by a flange and high-strength bolts. The arch base is pinned. Middle and edge arches are anchored at six base plates.

The edge arches of the hexagonal module have again the same dimensions as the 10 x 10-m square module for ease of connection.

Triangular module | The elements of the triangular module cover a base of 45 m^2; they can be set up alone, but are usually used in combination with the hexagonal or square modules as a modular building system to cover large areas. The technical details correspond to the ones of the other modules.

Membrane | The PVC-coated polyester fabric is translucent, UV-resistant and supplied with a flame-spread prevention treatment and with a dirt-repellent topcoat.

The steel edge cables are continuous and are enclosed in membrane sleeves; they are anchored by means of a threaded bar or a swaged thread fitting with nuts onto steel plates welded to the arch.

In the membrane corners a clamping plate is arranged on both sides of the membrane on top of a membrane reinforcing patch. The guy to the base plates or foundations consists of a tension tie of standard shackle, chain and turnbuckle tied to a steel connection plate, to which one valley cable each and two corner guys are connected.

Assembly

For the square module each arch is assembled from four parts, for ease of transport and erection; the parts are connected by a plate flange with high-strength bolts. The arches are assembled on the ground. After being fastened to the anchor plates with pins, they are turned upright and joined with diagonal struts. The membrane is connected to the arch and brought into its final shape and prestressed with the help of the valley cables. The arches can be erected without a hoist (i.e. using only a tirfor). Two arches joined by a diagonal compression strut are stable already, i.e. can stand alone unsupported.

The installation of the edge arches in the hexagonal module is similar to that of the square module. At the apex of the middle arches there is a twin pin connection in each quarter arch joining it to the central apex element; the upper pin acts as an installation joint when raising the two connected quarter arches. The lower bolt is brought in after the raising and then blocks the hinge, avoiding the formation of a kinematic link chain.

The triangular module is assembled similarly to the square module.

Accessories | There are variants for a raised installation with increased apex height, raised foundations and podia and up to 3-m high steel columns and purpose-made concrete columns. Rainwater is gathered in transparent polycarbonate elements at the base of the arches and drained away through rain pipes in the columns or under a simple installation floor.

The gap between abutting arches is sealed by a modular cover strip held in position with two light edge cables.

Detail view; Vierendeel arch with membrane guy and corner bracing by horizontal compression strut

Square module 10 x 10 m; 6 modules in front of an Italian palazzo

The side walls as space-enclosing elements may consist of coated fabric, glass or polycarbonate. Variants in apse form exist, and ones with planar, vertical walls with complete or partial space enclosure, with and without door openings. A wall with sliding glass doors has also been built.

Overall view of the excavation site at Desenzano

Roof over Archaeological Excavation Sites on Pianosa and in Desenzano sul Garda, Italy

Design

The excavation sites of the Roman villa Agrippinas on the island of Pianosa in the Tyrrhenian sea south of Elba are protected by a modular membrane roof with a space truss as primary structure. A similar system was used to protect the mosaics and wall remains of a villa in Desenzano sul Garda (at the south end of the Lago di Garda near Brescia), both with carefully designed details.

Brief

The roof of the excavation sites has the function of protecting the archaeological finds from the weather and simultaneously limiting the vegetation growth within the excavation site.

Desired were:

– Simple erection through prefabrication and application of a modular structure made from small and lightweight elements, being also easy to transport, to ease the sometimes difficult access to the excavation sites.

– A variable erection to allow a short-term adjustment during the survey of the foundations. The chosen structural system of the space truss is very flexible: If during the excavation of the foundations it turns out, that the planned supports must be moved because of the excavations, then a redistribution of the supports is possible by modifying the structural analysis and possibly by adding reinforcements, like for example additional cable stays.

– A later expansion of the excavations without changing the existing structure.

– No or only slight interference with the excavations, small foundations.

– A certain translucence was desired, to avoid too strong a shadow under the roof, which would have inhibited the excavations. A high translucence had to be avoided to protect the archaeological finds from UV- radiation, to avoid glare and to keep the vegetation low.

– Easy adaptability of the structure to uneven ground by variation of the column lengths.

Plan

The roof at Pianosa has an abundantly varied plan. The 4 x 4-m grid follows the outline of the excavation site and allows an extension of the structure in every direction in case of an expansion of the excavation area, without the need to change the existing structure. Roofing the

Client
Superintendentia Archeologica della Lombardia (for both projects).

Architects
General:
Pianosa: Architect Maurizio Giachetti; Collaborators/designing architects: Giovanna di Steffano, Saverio Innocenti
Desenzano sul Garda: Architect studio Albini-Helg-Piva, Milan
Membrane structure: Studio Tensoforma, Bergamo

Structural engineering
Membrane, structure, substructure: Studio Tensoforma, Bergamo

Contractor
Membrane and steel: Tensoforma Trading S.R.L

Completion
1990

Plan view of roof with space truss and membrane modules (Desenzano)

Interior view with sheet metal gutters and wind frames (Desenzano)

entire excavation site by a coherent membrane structure was not possible because of their special plan form.

In Desenzano sul Garda the roof consists of two near-rectangular plan shapes with 4 x 8 modules each, which are turned by 90° against one another. To adapt the plan form to the shape of the excavation site some edge modules were omitted.

Structure

The structure consists of tube columns and a space truss carrying the pyramid-shaped high point membrane elements. The steel structure is galvanised and painted.

The space truss serving as a roof structure is a double-layer grid of planar trusses with vertical compression struts and tension diagonal crosses. The tension diagonals consist of solid round steel bars with female threads and a two-piece, central tensioning ring at the crossing. The nodes are solid steel balls (Ø 90 mm) with milled planar bearing surfaces and bores with female thread to receive the bolts of the tension ties and compression struts. Conically turned steel parts with a shoulder/rebate form the transition between steel tube column and other compression members on one side and the bolt connection at the space truss ball node on the other. Tube columns are arranged on the grid lines along the edges and at re-entrant corners in the interior.

The three-dimensional structure was analysed electronically for the load cases of dead load, snow and wind.

Foundations, anchorage | The vertical column loads and the horizontal loads from the wind bracing are carried into concrete pad foundations. Due to the space truss roof structure acting as a horizontal diaphragm, it was not necessary to carry substantial tensile forces into the ground; no guys or tension anchors have been used. The horizontal forces from the membrane roof are transferred in the plane of the roof by the space truss and carried into the foundations by the wind bracing.

Bracing, tension members | In Pianosa the structure is braced laterally by a cross bracing with tension ties made of high-strength stainless steel cables with galvanised fittings, steel fixings and prestressing devices. In Desenzano the wind bracing consists of corner braces forming 'rigid' frame corners between edge columns and truss girders.

The tension diagonals of the wind bracing are spiral strands, with swaged eyelet fittings at the ends, connected to the tube columns with welded steel lugs and bolts.

The hangers for the high point suspension were originally conceived as round steel tension ties and were realised as a spiral strand with end thimbles connected at the high point with bolts between lug plates welded to the high point and at the upper chord nodes by means of ring bolts and pinned purpose-made fixings.

Below: Bottom chord node with tube column, sheet metal gutter and connection of truss members and membrane modules; above: upper chord node with truss member connection and hanger cables of the high point suspension, which simultaneously act as horizontal bracing of the space truss. 1:50
1 membrane module, 2 upper chord member, steel tube (Ø 88.9 mm), 3 bottom chord member, steel tube (Ø 88.9 mm), 4 vertical web member, steel tube (Ø 88.9 mm), 5 tension tie steel rod (Ø 20 mm), 6 node, solid steel ball (Ø 90 mm), 7 conical transition piece, turned steel part, 8 central connecting bolt in strut axis (not shown), 9 spacers, 10 upper connecting plate for high point hanger cables, 11 fixing for high point hanger cables, 12 connecting bolts, 13 high point hanger cables (here shown as tension rods, and simplified at the execution), 14 tie-member connection: threaded bar with hexagon, 15 lower connection plate for connection of the membrane corners, 16 tube column (Ø 137.9 x 5 mm), 17 conical transition piece, turned steel part, 18 sheet metal gutter, 19 washer

Space truss upper chord nodes with tension ties and compression struts, showing also the connection of the hanger cables of the membrane high points

Space truss bottom chord nodes with column and wind bracing connection

Membrane | The roof modules have a square plan and are fastened with their corners to the bottom chord nodes of the space truss. The high point is suspended from the upper chord nodes with stainless steel cables. The drainage is to the outside via sheet metal gutters arranged under the bottom chords and suspended from the truss nodes.

The membrane consists of PVC-coated polyester fabric (Ferrari Precontraint 702).

The sunlight is reflected to 70 %; the translucency is approx. 10 %, i.e. the membrane lets enough light through so that unacceptable shadows are avoided for the excavation work, but the membrane still protects the exhibits from direct exposure to the sun.

The high point structure comes in two parts. The membrane is fastened to a lower high-point ring through a steel clamping strip and a boltrope. The upper high-point ring is suspended with four hanger cables from the upper chord nodes of the space truss. Both rings are joined through an 'prestressing' bolt, whereby the membrane can be stressed and adjusted for manufacturing tolerances. As an alternative solution a modified construction with high points using a translucent light dome covering were also used.

At the edge the membrane is held through stainless steel edge cables in fabric sleeves. The rainwater is drained away via a non-structural membrane apron into the sheet metal gutters.

The two-piece membrane corners consist of steel plates and were clamped on with bolts above and below the membrane. The stainless steel edge cables were realised as 'endless' cables and anchored in eyelet plates at the node.

Construction

As happens frequently the details were modified at the shop drawing phase by the contractor and adjusted to his production facilities.

07.6 Diagonalenspannring und Membran-Hochpunkt mit transluzenter Abdeckung

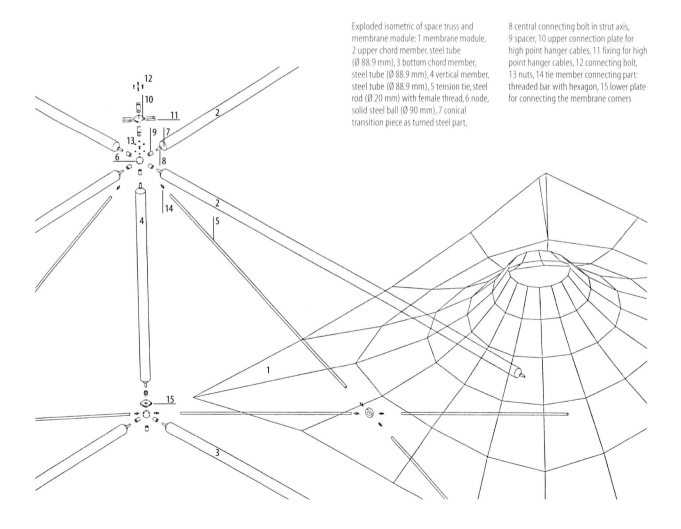

Exploded isometric of space truss and membrane module: 1 membrane module, 2 upper chord member, steel tube (Ø 88.9 mm), 3 bottom chord member, steel tube (Ø 88.9 mm), 4 vertical member, steel tube (Ø 88.9 mm), 5 tension tie, steel rod (Ø 20 mm) with female thread, 6 node, solid steel ball (Ø 90 mm), 7 conical transition piece as turned steel part, 8 central connecting bolt in strut axis, 9 spacer, 10 upper connection plate for high point hanger cables, 11 fixing for high point hanger cables, 12 connecting bolt, 13 nuts, 14 tie member connecting part: threaded bar with hexagon, 15 lower plate for connecting the membrane corners

Front view

Mobile Bandstand, New York, USA

Design

With an audience distributed over many different boroughs, New York City traditionally has used portable stages for outdoor summer concerts. The new, $3.4 mill. Carlos Moseley Music Pavilion – named after the chairman of the New York Philharmonic who, in 1965 introduced free summer concerts in the park – was to be used in the first summer for 30 concerts and performances in 16 parks in five boroughs. It is a newly developed, fully mechanised tensile structure, which is erected from five semi-trailers, two trucks and their loads and offers a stage large enough for a full orchestra and chorus. Provided there is firm ground and vehicle access, speaker towers, amplifier plant, stage and membrane roof can be set up within six hours. After a performance the 288 m^2 large open-air stage (276 m^2 of it under cover) can be disassembled and loaded in the same time, transported to the next location and to be set up again the following day.

To avoid obtaining special authorisation each time the stage is moved, weight and dimensions of the vehicles had to correspond to the US highway regulations, which stipulates a maximum height of 3.95 m and a length of 13.7 m for a tractor trailer bed and its cargo. Because of its rotating and hydraulically unfolding girders it also had to be approved by the Department for Cranes and Derricks, the authority responsible for movable buildings. Without permanent foundations it did not come, however, within the responsibility of the Department of Buildings.

The design strategy was to allow the shape and geometry of the structure to be completely determined through technology. The roof membrane received its form from the requirements for sound reflection and rain protection for the stage. The architectural quality was designed into the proportions and the interaction of the different elements. When the pavilion is illuminated at night, the structure disappears; only a hovering canopy remains visible, as a coloured backdrop for the performers on stage.

Technical installations, equipment

For images and subtitle projection an inflatable, circular membrane screen (Ø 4.8 m) is suspended above the membrane. The tripod trusses in front simultaneously serve as rigging for the theatrical lighting with fixed wiring; an additional aluminium truss for overhead lighting is suspended from a hanger cable, supporting also the membrane at mid-section.

Client
New York Philharmonic and the Metropolitan Opera and NYC Department of Cultural Affairs

Architects
FTL Architects plc

Structural engineering
Membrane: Buro Happold
Structure: M. G. McLaren

Specialist consultants
Lighting: Peter Wexler
Acoustics: Jaffe Acoustics

Contractors
General contractors + steel: Quickway Metal Fabricator
Membrane: Fabric Structures, Inc.

Completion
1991

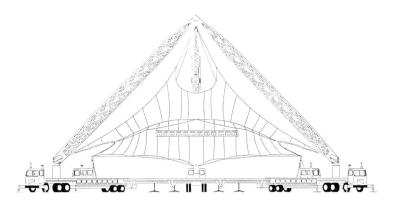

Front view 1:500

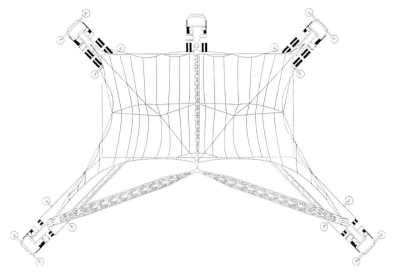

Plan 1:500

Acoustics | The audio system claims to produce outdoors the acoustics of an actual concert hall. The sound system was developed by Jaffe Acoustics exclusively for this project. The sound is transmitted via radio from an audio-visual control located 60 m from the stage to 24 speaker towers, set up in a semicircle of concentric rings in front of the stage. The battery-powered speakers, each on four collapsible legs are unfolded and raised telescopically to a height of 4.5 m and produce an almost ideal sound distribution and consistent sound level. The acousticians calculated and installed a delay of a fraction of a second in broadcasting music, corresponding to the time needed by the soundwaves to travel from the stage to the front speaker row. So the illusion is produced, that the amplified sound comes from the stage. The central speaker is the loudest, focusing attention on the stage. A slight delay in the sound transfer from the rear speakers simulates the reverberation of sound, which a listener would hear from the rear wall of an enclosed concert hall. Similarly the side speakers produce the effect of sound reflecting off the walls to the left and right.

In addition the saddle-shaped membrane, tied to the rear and the sides of the stage, reflects the music, and is supported in this by a semi-circular wall of aluminium reflector panels with plywood backing, standing on the stage.

Structure

Stage | The hinged stage floor (12.2 x 23.7 m) of eight marine plywood panels is carried by six lightweight aluminium beams, transported folded like an accordion and opening automatically with the aid of hydraulic pistons. The corners and the rear centre of the stage rest on a 21.3 m long and 35.9 m wide substructure set up beforehand.

Roof | The tripod trusses consist of a painted four-chord tubular truss; the two front trusses have a length of 26.4 m; a rear girder acts as a hydraulic hoist lifting the front trusses. The front girders are first unfolded to their full length; then fastened to the rear girder. As the rear girder unfolds with the aid of a hydraulic hinge, it raises the entire tripod to a maximum height of 20.7 m. Once fully elevated, a sliding steel pin locks the hinged truss into a stable tripod structure.

Foundations

Without requiring foundations or fixed anchorages to the ground, the trucks can be weighed with concrete ballast to counteract the thrust of tripod and tensile membrane. For the stage retractable foot pads are unfolded hydraulically and adjusted in height to provide suitable support points on uneven ground.

Roof membrane | The roof skin consists of a PVC-coated polyester fabric (Ferrari type 1002S with Fluortop T®) with a dirt-repellent PVDF-topcoat and it has a tensile strength of 3.92 kN/5 cm in warp and 3.80 kN/5 cm in fill direction. Due to large variation in erection the prestress of the membrane varies between 3.5 kN/m and 17.5 kN/m in warp and fill direction. The seams are high-frequency (HF-) welded.

The roof is assembled from six individual panels. Its 52 cutting pattern strips have a nominal width of 1.67 m. The total membrane surface is 339 m^2.

For maximum rain protection multiple edge cable scallops along the sides and the rear make it possible to draw the roof membrane and its corners as close to the stage edges as possible. Kevlar ropes (Ø19 mm) were used for the edge. Due to their flexibility and high strength they can remain in the edge cable pocket when the membrane is rolled up for storage, thus cutting at least 45 minutes from the otherwise required installation time, compared with the stiffer steel cables used for hangers and guys, which had to be re-inserted into the edge cable sleeves each time.

Aluminium corner plates and an additional membrane reinforcement at the corners serve as connection with the guy cables, anchoring the membrane in the structure. Man-made fibre webbing installed in pockets under the membrane and anchored at the corner plates reinforces the roof skin; this forms slight ridges in the membrane surface and so caters for sufficient curvature.

The Kevlar ropes will probably have to be replaced every two years, while the membrane has a service life of well up to twenty seasons depending on its handling by the construction crew, trained specially to raise and strike the pavilion.

Assembly

One truck transports the fabric membrane and lighting, a second one the electrical distribution and the stage equipment. One of five semitrailers carries the stage and the rear truss girder, two others the two front girders and a fourth the speaker towers. The fifth semitrailer contains folding beams, which open to provide the stage structure; from it hydraulic pistons unfold hinged panels, forming the stage surface.

The translucent membrane of the acoustics sail is transported rolled up in a protective ground sheet.

A forklift with special balloon tyres for minimal impact on the lawn places the twenty-four speaker towers around the site.

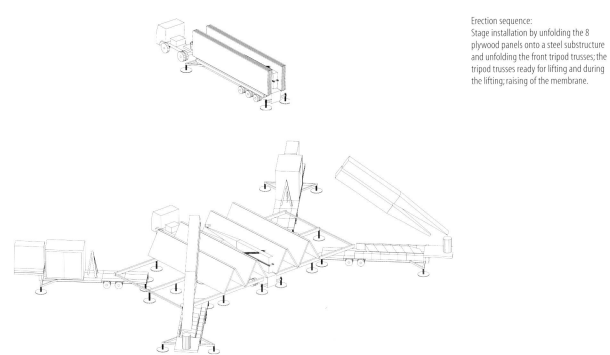

Erection sequence:
Stage installation by unfolding the 8 plywood panels onto a steel substructure and unfolding the front tripod trusses; the tripod trusses ready for lifting and during the lifting; raising of the membrane.

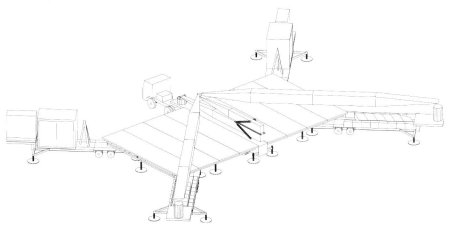

Bibliography

Library, Los Angeles
Architectural Review, 6/1993, p. 40 ff

Inland Revenue Centre, Nottingham
The Arup Journal, 4/1995, p. 11-17; Detail 6/1994, p. 785

Venafro, Pozilli
L'ARCA, No. 42, October 1990, p. 48-55 and No. 57, February 1992, p. 76-81; Building Design, No. 1017, 11. January 1991; Bouwen Met Staal, 105, March/April 1992, p. 19-24; Architectural Review, No. 1143, May 1992, p. 67-70; Le Tensostrutture a Membrana per l'architettura, 1993; Architektur Aktuell, No. 157, April 1993, p. 32-35; Space Design, No. 346, July 1993, p. 69-96; de Architect, March 1994, p. 80-87; H. C. Schulitz, Industriearchitektur in Europa, CONSTRUTEC PREIS 94, Berlin 1994; DBZ Deutsche Bauzeitschrift, No. 2/1995, p. 105-110; Architektur, May 1995, p. 30-35

Munich Zoo
Detail 4/1995, p. 628 ff; DR-Design Review Heft 9, September 1995, p. 22/9-23/9

High-Rise, Osoka
Keizo Sataka, Clouds and Water, in Fabrics & Architecture, November/ December 1993, Vol. 5, No. 6, p. 10-11

Leisure Centre, Riad
Interiors, March 1995, p. 62

Gymnasium, Tokio
Kazuo Ishii, Membrane Structures in Japan, SPS Publishing Company, 1995

Festival Theatre, Llangollen
International Eisteddfod Theatre, Llangollen, Year Book 1992; Architectural Review, Product Survey: Structures and cladding, 7/1993, p. 89

Mobile Theatre, Hamburg
Klaus Schmidt-Lorenz, Fliegende Bühne, in Design Report, 3/1995, p. 42-44; Die Buddy-Holly Story, in Sarnafil Membranbauten, January 1995; Die Sarnafil Buddy-Holly Story, in Sarna Hauszeitung 2/1994

'Heureka', Zurich
Imposante Holz-and Zeltkonstruktionen für die Heureka in Zürich, in Schweizer Baublatt, 37/38, 7. May 1991

Pneumatic Hall, Esslingen-Berkheim
Detail, 8/1996, p. 1204 and 1274; DB Deutsche Bauzeitung, 2/1996, p. 134; Blueprint, December 1996; Pro Architektur, 1/1997, p. 38 ff.; design report, 1/1997, p. 52-55; form, 1/1997

Expo Pavilion, Tsukuba
Kazuo Ishii, Membrane Structures in Japan, Tokio, 1995

Convention Centre, San Diego
Civil Engineering, November 1987, p. 44-46; Fabrics & Architecture, Winter 1989, p. 8-11 Architectural Record, August 1990, p. 55-62; Horst Berger, Light Structures, Structures of Light, Birkhäuser Publishers, Basle 1996, p. 134-141

Supermarket, Plymouth
Archetype, October 1994, p. 37; Leonardo, 1/1995, p. 53-55

Filling Station, Wanlin
Ph. Samyn, D. Melotte, M. Mollaert, Construction of Two Motorway Service Stations for FINA Europe, Asia-Pacific Conference on Shell and Spatial Structures, CCES-IASS APCS '96, May 1996, Bejing

Salzburg Airport

M. Rhomberg,"Von der Idee bis zur Realisierung. Die Planung der Membranbauten am Flughafen Salzburg", techtextil '95

Auditorium, Tarragona

Architectural Review, April 1994 (only ill.); Almanacco dell'Arca, 1994 (only ill.)

Climate-Control Parasols, Medina

Techtextil-Telegramm, 24. January 1994, No. 31; Fabrics & Architecture, May/June 1994, p. 20 ff
Techtextil 4.2, 6. Internationales Techtextil Symposium 1994, Beitrag No. 4.23; db deutsche bauzeitung 9/93, p. 68; L'ARCA, July/August 1993, No. 73, p. 10 ff; Frei Otto, Bodo Rasch: Gestalt finden, Edition Axel Menges, 1995

Excavation Sites, Pianosa, Desenzano

L'ARCA, July 1993; Detail, 6/1994
MODULO, October 1994

Mobile Bandstand, New York

Architecture, September 1991, p. 102-105; Interiors, March 1995, p. 56

Acknowledgements

All illustrations, except those by the author as mentioned below, are courtesy of the architects, engineers and clients.

Atelier One 87 t, 89, 91; Horst Berger 128, 129 b, 130 l; Light Structures/Horst Berger 7, 130 b, 131 l; Horst Berger & Partner 129 t, 131 r; Stefano Bertino, Tensoforma 78, 79u, 80, 81 t+m, 160, 162, 163 b, 164, 165, 166 m+b, 167 t; Chorley & Handford 40 tl; Brian Dent, Godwin Austen Johnson 71, 74; Horst Dürr/IF 11l; Festo KG 22-24, 102-105; FTL/Happold 75-77; FTL Associates 169, 171 t; Graham Gaunt/Ove Arup & Partners 39, 40 tr, 41, 42 t; Dennis Gilbert 101, 132 t, 113, 134 tr; Godwin Austen Johnson 72; Jeff Goldberg/Esto 168, 171 b; Craig Hodgetts 32, 33, 34 t, 36; Hodgetts + Fung, Architects 34 b, 35; IF Dürr 54-57, 111-113, 114-117, 141-146; IF Ingenieurgemeinschaft Flächentragwerke 13l, 15l; IPL-Ingenieurplanung Leichtbau 13 r, 49, 51-53, 93 b, 96, 98 l, 99, 101 tl; Kajima Design 82-85; Yasufumi Kijima, Keikaku Kankio Kenchiku, Design Office 25; Angelo Kaunat 31, 62, 64, 65; AB Kochta 63, 67 t; Friedemann Kugel 60 l, 61 t; Richard Leeney, D.Y. Davies Associates 19, 86, 88, 90; Eduard Lehner 66; F. Loze & Archipress Paris 26 b; Ken Naverson 37; Ove Arup & Partners 13 b, 40 b, 42 m+b, 125 t, 127 tl, 134 l+m, 135; Permafab Pty Limited 127 tr+b; Benoit Pesle 26 t, 58, 59 ol, 60 r; Matteo Piazza 27-30, 43, 44 t, 45, 47; C. Putler, R. Armiger 132 b; Rhomberg 142 t; Rubb Building Systems 38; Samyn et Associés 44 b, 46, 137, 140; Samyn & Partners, Ch. Bastin & J. Evrard 136, 138, 139; Sarnafil AG 18 t, 92, 95, 97, 98 t; Keizo Sataka 106, 108 t; Keizo Sataka, Institute of International Environment 68-70, 107, 108 b, 119-123; Schneider-Zimmerhackl 49 t+m, 50 t; Hans-Joachim Schock 7, 8, 9 l, 10 r, 11 r, 14 l, 15 r, 67, 73 b, 87 m+b, 101 tr+m+b, 109, 110, 163 t; SL Sonderkonstruktionen & Leichtbau GmbH 17, 153-157; Stahlbau Zwickau 18 b, 48, 50 b; Tensoforma 14 r, 20/21, 79 t+m, 81 b, 147-152, 161, 166 t, 167 b; Urban Project GmbH 93 t, 94; Jörg Wagner 73 t; Colin Wade, Ove Arup & Partners 124, 125 r, 126; Walter Zbinden Atelier d'Architecture et d'Ingéniérie 59 tr+m+b

Index

AIC Architekten und Ingenieure Coop 160
Airport, Salzburg 15, 141
Al Naboodah Contracting 71
Al Naboodah Laing 71
Albini-Helg-Piva 164
Alico 71
American Constructors California, Inc. 33
Angst-Obi, P. 97
ARGE Hochtief - Oevermann - Groh 48
Atelier One 86

Baltensberger 97
Bergbauer 62
Berger, Horst & Partners 128
Berlin 54
Bertino, Stefano 160
Big Wave 119
Birdair 75, 128
Birse Construction 132
Blum 114
Bond, James, Norrie, Marsden 124
Buddy Holly Musical 92
Buro Happold 75, 153, 168
Butterfly House, Berlin 54

CAMP DE MART, Tarragona 14, 20, 147
Canobbio spa 18, 27-30, 43, 48
Carnivore and Palm House, Munich 31, 62
CERCON 75
Climate-Control Parasols, Medina 17, 153
Clyde Canvas Goods & Structures Ltd. 86
Colux 111
Convention Centre, San Diego 128
Davies Associates 86
Davis Langdon & Seah 124
De Boer Tenten nv 136
Deems Lewis McKinley 128
Dixon, Jeremy 132
Dove Associates 71
Dragages et Travaux Publics 124
DSB Deutsche Schlauchboot 102
Dubai 71
Dürr, Horst 54, 111, 141

Equation Lighting Ltd. 86
Ericson, Arthur & Partners 128

Esslingen-Berkheim 22, 24, 102

Fabric Structures, Inc. 168
Feinberg, Michael Patrick Byrne & Associates 33
Ferroglas 141
Festival Theatre, Llangollen 19, 86
Festo KG 2, 22-24, 102
Flontex GmbH 54
FTL Architects plc 168
FTL/Happold 75

Generer and Partners 86
Gerritz 48
Giachetti, Maurizio 164
Glasbau Seele 62
Gobiet sa 136
Godwin Austen Johnson 71
Gorle 78
Government Stadium, Hong Kong 124
Gutierrez, Julio Cadena 147

Hamburg 92
Hellmuth, Obata & Kassabaum, Inc., Sports Facilities Group 124
Hennicke, Jürgen 160
Herbrich, R. 141
Herrschmann, Dieter 62
Hervé 58
'Heureka' 18, 97
High-Rise Facade, Osaka 68
Hiroshima 119
Hodgetts & Fung Design 33
Hong Kong 124
Hopkins, Michael & Partners, London 39

IBE Rolf Günther 62
IF Ingenieurgemeinschaft Flächentragwerke 54, 111, 114, 141
Inland Revenue Centre 39
Institute of International Environment 68, 106, 119
Invernizzi, F. 43
IPL - Ingenieurplanung Leichtbau 43, 48, 58, 62, 92, 97

Jaffe Acoustics 168
Jaschek 102
Johnson, Brian 71
Johnson, Spurier 128

Jones, Edward 132
Jurina, Lorenzo 147

Kaetsu Memorial Gymnasium 82
Kajima Design 82
Kamal, Ismail 153
Kiefer, Michael 48
Koch 102
Kochta, Herbert 62
KOIT High-tex GmbH 39, 62, 102, 153
Krantz 153
Kugel, Friedeman 58

Laing Management 39
Landrell Fabric Engineering Ltd. 132
Lehner, E. 62
Library at UCLA 32, 33
Liebherr-Werk Ehingen GmbH 153
Llangollen 19, 86
Los Angeles 32, 33
Loschky, Marquart & Nesholm 128
Lycée la Bruyère, Versailles 58

Manning, Martin 132
Martin, A. C. & Associates 33
Martin, John A. & Associates 128
Martínez I Lanzas, Angel 147
Masserberg 18, 48
Matsui, Gengo 119
McLaren 168
MEDIADROM Mobile Air Hall 11, 114
Medina 17, 153
Merlini e Natalini 78
Ministerial Leisure Centre, Riyadh 75
Mobile Theatre, Hamburg 92
Mobile Bandstand, New York 168
Mobile Building System AIC-Tensoforma 160
Mollaert, M. 136
Möller, Eberhardt + Partner 48
Morique Building Co. 68
Moureau sa 136
Munich 31, 62

New York 168
Nolte, Carl GmbH & Co 141
Nolting KG 114
Nomura Co., Ltd. 119
Nottingham 39

Ollé I Ribé, Josep 147
Orio Gymnasium 25
Osaka 68
Ove Arup & Partners 39, 124, 132

Permafab Pty Limited 124
Pfeifer Seil- und Hebetechnik GmbH & Co 75
Philipp Holzmann AG 92
Pianosa and Desenzano sul Garda 164
Plüss + Meyer 97
Plymouth 132
Pneumatic Hall 'Airtecture' 22, 24, 102
Polynederlands BV 114
Pozzilli 27, 28, 30, 43

Quickway Metal Fabricator 168
Quigley, Patrick B. & Associates, Inc. 33

Rasch, Bodo 153
Rathgeb, Willi 102
Research Laboratory, Venafro 27, 28, 30, 43
Rhomberg, Michael 141
Rice, Peter 132
Riyadh 75
Robert Englekirk, Inc. 33
Robert Matthew Johnson Marshall 71
Rodio S.A. 147
Rowan, Williams, Davies & Irwin Inc. 124
Rubb Building Systems, Inc. 33

S + H Nolting 92
Sainsbury's Supermarket, Plymouth 10, 132
Saito, Masao 106
Salzburg 141
Samyn et Associés S.P.R.L. 43, 136
San Diego 128
Sarnafil AG 92
Sarnatent, Sarna Kunststoff AG 97
Sataka, Keizo 68, 106, 119
Saudi Binladin Group 153
Schaupp 62
Schlaich, Bergermann + Partner 62
Schmidt, Wolfgang Nikolaus 111
Schneider-Zimmerhackl, Laurens 48
Schock, Hans-Joachim 71
Schraepel 114
Schröder, Paul 48
Swimming Pool, Ettlingen-Schöllbronn 9
Swimming Pool Miramar, Weinheim 10
Sellhorn Ingenieurgesellschaft 92
Service Station, Wanlin 26, 136
SETESCO 43, 136
Sheetfabs Nottingham Ltd. 86
SICOS 43
Simmoneau 71
Simon Construction Co., Ltd. 68
Simpson, Adrienne 71
SL Sonderkonstruktionen und Leichtbau GmbH 153

175

Stahlbau Heinrich Weller 48
Stahlbau Zwickau 48
Starrs 58
Stendtke + Vedder 48
Straub, Johannes Peter 97
Stromeyer & Wagner GmbH 71, 111
Stromeyer Ingenieurbau 58
Studio H 43
Studio Tensoforma 78, 147, 160, 164
Sullivan Partnership 33
Sun Corporation 68
Suntory Ltd 106
Suntory Pavilion, Tsukuba 106
Synergy 86
Syska & Hennesey 128

T.I.S & Partners 106
Taisei Corp. 68
Taiyo Kogyo Corp 82, 106, 119
Takenaka Corp. 106
Tarragona 20, 147
Teachers Training Collge, Ludwigsburg 9
Technet Gmbh 136
Tennis Court, Gorle 78
Tennis Hall, Essen 8
Tensoforma Trading 147, 160, 164
Teutsch + Partner 62
Thomas Thompson Lighting Design 75

Tilbury Douglas Construction Ltd. 86
Tokio 82
Trade Fair Stand Automechanika 111
Tsukuba 106
Tubemasters 75
Turner and Townsend Quantity Surveyors 39
Tuto - Saliba - Perini 128

Urban Project GmbH 92

Venafro 27, 28, 30, 43
Venning Hope 86
Versailles 58
Vistawall Architectural Products 75
Vrije Universiteit Brussel 136

Wanlin 26, 136
Westbury Tubular Structures Ltd. 86, 132
Wexler, Peter 168
WNS Wolfgang Nikolaus Schmidt Design GmbH 111, 114
Woodward – Clyde Consultants 128
Yacht Club, Dubai 71

Yulura Hodiday Resort, Alice Springs 8

Zbinden, Walter 58
Zinner & Sohn 62
Zurich 18, 97